Events of future history will likely be of the same nature as the history of the past, as long as men are men.

— Thucydides, 400BCE

Many men go fishing all of their lives without knowing that it is not fish they are after.

— Henry David Thoreau, 1853

Science, my lad, is made up of many mistakes, but they are mistakes which is it useful to make, because they lead little by little to the truth.

— Jules Verne, *From the Earth to the Moon*, 1864

It's in the nature of the human being. We're required to do these things just as salmon swim upstream.

— Neil Armstrong, 1969

PRAISE

For Steve Harris

Harris aims to be more than simply a teller of history. He also wants us to learn from history.

— *Weekend Australian*

Under the Influence of Salmon: How a Man and a Fish Turned our World Upside Down (2025)

If you enjoy stories of obsession, then this is a book for you. In recounting one man's lifelong quest to introduce salmon to Tasmania, Harris carries the reader to a very different era from our own. Meticulous attention to detail and clear prose make it a compulsively good read.

— Anders Halverson. Aquatic ecologist, journalist and author of *An Entirely Synthetic Fish: How Rainbow Trout Beguiled America and Overran the World*

In this spellbinding, ripping account of an overlooked saga from Australia's history, Steve Harris brilliantly captures the heroic folly of a group of dreamers – single-minded men hellbent on bringing the Old World's most famous sport fish to the Antipodes. If fish and fishing mean half as much to you as they do to me, I doubt you'll be able to let this one go!

— Steve 'Starlo' Starling. Fishing Communicator & Educator

An inspiring and detailed narrative that delves into the nuances of the original idea and the many people involved with bringing salmon, and ultimately trout to the Southern Hemisphere.

— Gary France. *Trout Guide*, Cressy, Tasmania

As an angler, scientist and former fisheries administrator I am not sure which part of me enjoyed this fascinating book the most, but I was well and truly hooked from the start.

— Rob Sloane. Founding Editor, *FlyLife Magazine*

The Lost Boys of Mr Dickens (2019)

History as suspenseful, affecting and moving as the best films and fiction.

— Robert Drewe

Offers so much more than you'd expect ... profound, original ... the book entertains and instructs.

— Miriam Margolyes

It's a tragic tale that reads like it's straight out of a Charles Dickens novel.

— *Herald Sun*

Enthralling ... this absorbing book leaves the reader in little doubt that (the boys) would have been permanently scarred by British justice.

— *Weekend Australian*

Harris skilfully portrays one of the most sobering and saddest stories in Australian history.

— Geoffrey Blainey

A riveting account of how these boys were no longer seen as children.

— Dickens Society (NSW)

Brings to life a most bizarre social experiment and all its grotesqueness in engrossing form. A service to Australian and British readers.
— Tom Keneally

Lost Boys may be a history, but it reads like a novel ... and an excellent one.
— Meg Keneally

A moving story ... embedded in a penetrating and thorough study of a grim part of Tasmania's history.
— Alison Alexander

This book makes an important historical contribution ... truly engaging, a great read.
— *Journal of Law and History*

An absolutely wonderful work of scholarship.
— Pete Hay

The Prince and the Assassin (2017)

Marvellous, a truly gripping tale, wonderfully researched.
— Jane Ridley. Professor of Modern History, University of Buckingham

Harris has excavated from the past a little remembered incident that had huge implications at the time and still resonates today. It has global resonance.
— *Irish Times*

An enthralling tale ... Harris constructs a powerful account of a disgraceful episode in British colonial history.
— *Dickensian* (London)

Engrossing...reads like a gripping historical courtroom thriller.
— *Sydney Morning Herald*

A fascinating but forgotten story ... resonates powerful with ongoing debates about youth crime, responsibility and reform.
— Dr Matthew Allen. Historian/criminologist, University of New England.

History, politics, religion and the law are woven into an entertaining story of empire loyalty and disloyalty.
— *Law Institute Victoria Journal*

Harris skilfully weaves together the lives of these two men and explores the impact that their encounter had on Australia.
— Centre for Independent Studies

Solomon's Noose (2015)

It's up there as a classic. Harris is one of our better popular historians.
— Adam Courtenay

Intriguing story ... captivating tale ... a fascinating read.
— *The Australian*

Impressive research, a story that challenges the imagination, except that it's true.
— Les Carlyon

The reader [will be] indisputably captivated ... moving and often poignant.
— *Tasmanian Historical Research Association*

Published by Melbourne Books
Level 9, 100 Collins Street,
Melbourne, VIC 3000
Australia
www.melbournebooks.com.au
info@melbournebooks.com.au

Title: Under the Influence of Salmon: How a Man and a Fish Turned Our World Upside Down
Author: Steve Harris
ISBN: 9781922779472
Publisher: David Tenenbaum
Designer: Holly Lambert

NATIONAL LIBRARY OF AUSTRALIA

A catalogue record for this book is available from the National Library of Australia

UNDER THE INFLUENCE OF SALMON

OF SALMON

How a Man and a Fish Turned Our World
Upside Down

M
Melbourne Books

Steve Harris

CONTENTS

To Maureen, for her love and support.

FOREWORD
GEOFFREY BLAINEY, AC

In the 1860s, James Youl conducted one of the most difficult, long-distance experiments so far attempted in the world. Originally a Tasmanian sheep grazier, he was not really prepared for these experiments; that makes him even more remarkable. This absorbing book brings to life his adventures and even their roundabout influence on what we eat today.

For several centuries, wealthy people in the British countryside enjoyed salmon fishing as a happy pursuit. Many were inspired by *The Compleat Angler* by Izaak Walton, one of the most quotable and reprinted books in the English language. When Youl settled down in England, however, the salmon was no longer Walton's anointed 'monarch of the rivers'. Through overfishing and industrial and urban pollution, the beloved salmon were becoming scarce in their natural waters of Britain, Europe and North America.

Eager attempts were made to replenish the streams with salmon and their eggs or ova. Perhaps through new friendships, Youl joined the crusade. He inspected rivers in England, Ireland and France; he pestered naturalists and scientists; and he became more ambitious: 'I thought what a grand thing it would be to introduce this noble fish into the Australian and New Zealand rivers and was determined to use every effort in my power.' His spirits leaped when he heard that fish hatcheries in Norway and Ireland were selling quantities of fresh salmon in local street markets – all the result of salmon released into rivers where 'they were previously unknown'. Youl resolved to achieve what was deemed impossible and unimaginable: to breed salmon, and then trout, in the Southern Hemisphere for the first time.

Steve Harris' telling of this evocative story involved heavy research, for Youl – who died in London in 1904 – seems to have bequeathed no diaries and few private letters. But he did leave behind a remarkable project: Youl would be overjoyed to know that today numerous waters in his own Tasmania – including the D'Entrecasteaux Channel, Storm Bay, Huon Estuary and Macquarie Harbour – are devoted to the breeding of farmed 'salmon'. In terms of weight, this contemporary impostor now dominates supermarket fish shelves and fish restaurants in Australia.

Youl also seeded today's often vociferous debates: about the future of wild salmon; the costs and benefits of farmed salmon as food for the inexorably rising world population; humankind's relentless harvesting of Nature; and the balancing of new technology and profits against the existential need for a healthy and sustainable planet.

But – as this book emphasises – it is in the human spirit to explore, pursue and meddle.

His book reminds us that history is often most revealing when looked at through the eyes and mindsets of ordinary men and women. As much as the leader of empires, leaders such as Youl re-imagine, reshape and expand the art of knowledge.

As a journalist in Melbourne, Harris was successively editor-in-chief of two of the nation's largest daily newspapers. *Under the Influence of Salmon*, written in his next career as a historian, gains from his observant eye for detail, his willingness to fish and fish again for new evidence, his patience in explaining complicated swings of zoology and ichthyology, and his tenacity when the research swung in unexpected directions. Affection for his native Tasmania pervades this valuable book.

INTRODUCTION

Few who see its sleek, silvery, streamlined shape can come away without feeling they have probably seen Nature's most beautifully designed species. Or witness it leaping three metres out of the water without admiring its physical capacity, bravery, and will. Or pursue it with a rod without reflecting on the beauty of Nature's stage and the fine line between life and death.

It has been ever thus. Humans and animals have always shared a special relationship, but of more than two million known species, one is numero uno – the one which connects more, on more levels, than any other.

Its name is salmon, Atlantic salmon, *Salmo salar. For* centuries the most admired for its aspirational human qualities of beauty, athleticism, courage against the odds and a fierce determination to ensure survival of the species. A powerful symbol of life, beauty, divinity and mythology since cave days, the silent salmon has spoken volumes in the narrative of civilisations across the globe, been a lead actor in the ever-evolving interplay of humans and Nature, and an existential barometer of the world all species share.

Paleolithic relief sculpture of salmon in a cave in the Vézère Valley of the French Dordogne region, dated 23,000BCE. Photograph © Wellcome Collection.

For a long time, my own connection with salmon had not extended much beyond boyhood memories of pursuing so-called 'cocky salmon' (not then knowing it was a misnomer, not a true salmon at all) with my dad in the Leven River on the north-west coast of Tasmania, and the adult pleasures of salmon steaks, sushi and sashimi. That all changed when, on some desktop trawling on another book project, I caught an intriguing reference to something which happened a century and a half ago: a future King of England planning to visit salmon ponds in then Van Diemen's Land.

The tyrant of curiosity took hold. How could there have been salmon ponds anywhere in the Southern Hemisphere when the fish had then only ever existed in the Northern Hemisphere? Had the fish miraculously swum across the globe? Or had someone turned Nature on its head by transplanting the fish across 16,000 miles (25,700 kilometres)[1] of dangerous oceans and the tropics to live in alien waters at the bottom of the world? Who would think of such a thing? And why? And how would anyone actually do that? Did my homeland, one of the British Empire's most remote settlements, and a convict colony to boot, have an obsession with salmon? And how might this connect with today's controversies about the possible extinction of wild salmon, the

---

1   Imperial and Fahrenheit measures, as were used contemporaneously, have been utilised, with metric conversions discretionally provided.

booming and controversial and controversial salmon farming industry, and debates about how to feed a world population rising to 10 billion?

Emerging answers unveiled a tale as unimaginable as those then told in Jules Verne's fantastical Voyages Extraordinaires series, including *Journey to the Centre of the Earth* and *From the Earth to the Moon*. The notion of a few men to have the world's most admired fish swim, for the first time, on the other side of the world was seen as unimaginable as people not only reaching the moon, but fishing on it.

It was widely dismissed as madness, but some thought it just mad not to try. One was James Arndell Youl, the son of a preacher who set about challenging the laws of God and the known natural sciences to turn salmon, and the world, upside down.

With the aid of salmon devotees on both sides of the world, Youl's extraordinary efforts resulted in the most audacious, romantic, and intensive wildlife feat the world has ever seen. His quest captured the imagination of the world, drawing in giants of literature, science, royalty, education, government, and conservation, including Charles Darwin, Charles Dickens, Queen Victoria, Napoleon, Ulysses Grant, Garibaldi, Thomas Huxley, Louis Agassiz, Anthony Trollope and George Marsh – all seduced, influenced or bewitched by salmon.

This book endeavours to weave biographical threads: an extraordinary fish which mirrors human desires and values; a unique convict island desperate to find a new identity; a remarkable man consumed by an ambition to be a 'Mr So and So'. It goes to identity and place, and belonging, of man and fish. And it reveals the seeds of today's debates about global food security, biotechnological revolution, industrialised salmon farming, and environmental sustainability.

In the prism of the 'then' which has seeded the 'now' and is the genesis of the 'next', one hears the constancy of voices of discovery and ambition, identity and destiny, risk and reward, human nature and Nature, and 'madness'.

Perhaps, like me, after looking into the prism and hearing those voices, you will not look at salmon in quite the same way again.

1

SEEDS OF SEDUCTION

In the deep basement of time, ancestral salmon swam long before man and woman walked the land, the safe place for hunting and gathering. Life beneath the surface of the water was long one of silence and mystery.

So, when humanity first witnessed a salmon leap from the water into the sky, it was as wondrous as the brightest star. This was immediately and eternally special. The lion might be king of the jungle, and the stag the monarch of the glen, but salmon became king of the kings. A species which spoke to humanity's inner soul by exhibiting admired human traits: beauty, in its exquisite design; athleticism, in its ability to fly two or three metres into the air; bravery, in facing life-death moments; strength, by swimming against the tide; resilience, by never backing down or giving up; conviction, by leaps of faith. And, to seal the deal, a delicious meal.

Since Palaeolithic humans first scratched images of 'hero' salmon on cave walls, salmon was progressively eulogised and mythologised

Perhaps first depiction of fishing with a rod, from tomb of Beni Hasan. 2000–1650BCE, Egypt.

in ancient kingdoms, empires and Indigenous communities across Mesopotamian, Assyrian, Roman, Gaul, Chinese, Japanese, Russian, Druid, Pict, Celtic, Slavic, Baltic, Hellenic, and North American worlds. Each came to be seduced, anointing salmon as the first global superstar, an icon admired and revered for its beauty and as the source of divinity and wisdom. Recognised by Christian, Buddhist and Hindu religions, salmon was immortalised in stone and bone carvings, papyrus, jewellery, mosaics, fables, myths, religious ceremonies, poetry, literature and art.

But while humanity loved salmon, it did not know salmon, just that it fascinated, amazed, mystified and betwitched. They did not yet know that millions of years of climate changes and ice ages forced ancestral salmon to adapt to Earth's changing shapes and levels of land, sea, rivers and lakes. Or whether the original fish were of salt or fresh water or that descendants were among the one percent of fish species which could live in both. Or appreciate the salmon's extraordinary changes in physiology and colour masking its identity. Or understand how on earth it could find its migratory way into open seas and back to its original river nursery.

Salmon was given its scientific name, *Salmo salar*, in 1758 when the father of modern taxonomy, Swedish biologist Carl Linnaeus, married Roman and Gaul terms for salmon and leaper. But there was not much scientific knowledge behind the name.

First Peoples around the world were content with what they understood: as Nature sustained them, so in kinship they needed to take and share only what they needed. But 200 years ago, 'civilised' humans began to re-imagine the world as one in which they were 'homo deus', supreme gods of all species, with all providences of Nature subservient to a relentless appetite to conquer, domesticate, manipulate, harvest, and exploit.

With fish seen to be an endless source of profit and food, salmon was loved to death. Excessive and indiscriminate fishing disrupted its lifecycle, and agricultural-industrial-urban revolutions in the nineteenth century reshaped and polluted the landscape and waterways, pushing salmon to the brink of extermination in many parts of Europe, Britain and North America.

The Europeans' supreme faith that minds and technology could conquer all was blind to the price of greed, ignorance and stupidity. Charles Dickens gave voice to those fretting over the salmon's survival: 'A few years, a little more over-population, a few more tons of factory poisons, a few fresh poaching devices ... the salmon will be done ... he will be extinct.' As American naturalist-philosopher Henry David Thoreau asked: 'Who hears the fishes when they cry?'

But while those in the Northern Hemisphere were fighting salmon wars over the fish's very identity, lifecycle, and future, those in the Southern Hemisphere wanted the world to hear their very different cry: Why in God's name did He exhort humanity to 'rule over the fish' but confine His most admired species to the Northern Hemisphere?

Why salmon was only found in the Old World was part of the riddle that was salmon. Many dismissed it as God's way – 'if he wanted salmon in the south, He would have put them there' – but others in the Antipodes, or opposite, of the known world, had different ideas.

One was the son of a preacher from the island of Van Diemen's Land, one of the British Empire's most remote colonies, born of its attempt to seed a 'new' country by 'draining the swamp' of its own unwanted men, women and children. James Arndell Youl was not a convict, nor one with any affinity for fish or fishing. But he developed a dream to have salmon swim for the first time in the Southern Hemisphere, specifically the island at the bottom of the Great Southern Land.

It was a dream 40 years in the making. As a boy, not yet 10, he was dispatched in 1819 from his family amid its move from the rough

Map of Van Dieman's Land, c. nineteenth century.

hinterland of Sydney to even rougher Van Diemen's Land. Rev John Youl and his wife Jane wanted their eldest to be educated in the 'Mother Country' and perhaps follow a preacher's path.

But after a period in a private school in Essex – perhaps near family members who, like Rev Youl's ancestors, had migrated from Scotland to England – the windows of James' mind were opened by what he came to call 'that great Babylon, London'. At a pioneering branch of the Mechanics Institute in Chancery Lane, Holborn, ecclesiastical history was offered – but he was more drawn to Dr George Birkbeck's preaching about the new spirit of the 'steam intellect' age, where 'knowledge is power', and 'the wisdom of our ancestors' was to be eschewed. Just look, Birkbeck said, at an ordinary man's development of the steam engine: understand that, like the biblical beggar Lazarus, no man must be content 'with the crumbs which fall from the rich man's store'.

Another window was was a fishmonger's shop. Youl was mesmerised by a magnificent fish, a metre long, holding centre stage on a shining marble slab, surrounded by lesser species amid huge blocks of Scandinavian ice, ferns and palms, and crystal bowls of goldfish. 'I would never forget,' he later wrote, 'seeing princely salmon, that beautiful fish I so often admired on the marble slabs.'

He carried such imprints when he sailed to Van Diemen's Land in 1825 for a family Christmas and his 14th birthday. Just proclaimed the second colony, separate from New South Wales, the island was more than 16,000 miles (25,700 kilometres) separate from Babylon – roughly half the size of England and much closer to Antarctica.

Youl returned at the midpoint of a half century of transportation of nearly 80,000 convicts being banished 'beyond the seas', out of sight and mind. Convicts comprised some 60% of the population, but the whole island was run as a prison: its 'walls' were endless oceans to the east, west and south, and treacherous straits to the north. Only for secondary offences were convicts held in penitentiaries like Port Arthur or assigned to chain gangs, while gentry settlers became keepers of slave-like assignees. Rum was the de facto currency, even 'respectable' men threatened to resolve disputes by duelling, and newspapers dined out on tales of violence, sodomy and cannibalism.

It was the ugly side of a new world being built from the old, a new order imposed on the originals. After walking the southern plains of the Australian continent for 40,000 years before an ice age melt created the island, Aboriginal communities were wilting under the weight of white man's germs, guns and poison, and judgement that 'inferior savages' were no barrier to imperial and personal appetites.

Arriving in Hobart Town[1], Youl took in English-named inns (one for every 200 people), English-style cottages, English fashion, and English sandstone buildings. Then a coach with Royal Arms and scarlet-jacketed guards took him on 120 miles (190 kilometres) of convict-made road through staging posts with familiar names like Brighton and Ross.

Past and future was evident near the end of the 15-hour journey, just past Epping Forest, at Snake Banks. With two old Aboriginal men watching warily from the distance, the western side of the road featured

---

1. 1   Hobart Town then became Hobarton, then Hobart in 1842.

Hobart Town, 1820s. Photograph of sketch by Alan Carswell, 1821. Courtesy University of Tasmania Special and Rare Collections.

a convict chain gang – a common sight described by one visitor as 'like an ugly nose (which) spoils the face of the country'. On the eastern side lay what William Paterson, a Scottish penal commandant and Rev Youl associate, described as 'the most superior tract of arable and grazing land I have witnessed ... one of the most beautiful countries of the world.'

At beautiful Symmons Plains, named after three Symmons brothers who previously held grazing rights, Youl reunited with his preacher father, mother and three siblings he barely remembered, and met four new brothers and sisters. Reflecting the love of 'home', Rev Youl's land grant was in the Scottish-sounding Morven district (now Evandale) near the Yorkshire-sounding South Esk, flowing from the foot of Scottish-sounding Ben Lomond, into the Cornish-sounding Tamar estuary and Launceston, Australia's third settlement.

But it was plain Youl might not return to London. His father had never been of robust strength – it was family lore that he only survived cannibalism as a missionary in Tahiti because he was too thin – and from his original post at Port Dalrymple (now George Town) he struggled with health issues and the strains of what he called a 'spiritual darkness'. Van Diemen's Land's air was the freshest in the world, but few breathed easily. Whalers declared: 'Below 40 degrees south there is no law, below

50 degrees there is no God.' With Governor George Arthur's arbitrary law making everyone subservient to 'prison' rule and surveillance in the world's first police state, justice and fairness were elusive.

God's work fell to just one preacher in the south, and Rev Youl in the north, where he detailed 'wickedness, blasphemy, drunkenness, adultery and vice of every description'. When capital judgment weighed, as it did proportionally more than anywhere else in the world, it was his heavy duty to minister condemned men and women, young and old, and walk them to the scaffold.

While he still managed plantings in what had been a desert – the town's first substantial school, the Cornwall Collegiate Institution, and first church, St John's – the son could see the obvious toll on his father and stayed on. After another cold, wet winter fighting asthma and trekking the north for weddings, burials and preaching (and fighting asthma), Rev Youl faced a harrowing few weeks. In early 1827, he had to minister all night to six distressed men waiting to be hanged side by side like tassels on a cord, followed soon after by another group of eight. Then the well-connected *Colonial Times* in Hobart reported: 'The Rev. Youl is to be removed from his flock' to a minor role at The Punt, where a ferry across the South Esk served convicts, military and a small number of settlers. He and his family were shocked, and Launceston newspapers reported he was 'heard to say (it) would break his heart ... (he) heard his doom and wept'. Within days he was dead, aged 53 but looking 'near seventy'.

James Youl was just 17. The *Colonial Times* reported 'the deceased's eldest son ... left (Hobart) in a chaise for the house of mourning, seriously indisposed.' A family friend suggested he 'was not expected to survive his father's funeral, from grief'. Youl had possibly been at the British Hotel where a crowded meeting formed Australia's first Mechanics Institution to provide the 'instructive pleasure' he had enjoyed in London. With his distressed mother Jane within weeks of delivering a ninth child, Youl's grief was undoubtedly compounded by needing to bury thoughts of returning.

Then he burned with anger. The ambitious Governor George Arthur had taken 'umbrage' over Rev Youl's plain-speaking reports of the 'alarming immorality' on his watch, and was empowered by his friend, autocratic archdeacon Thomas Scott, the Oxford-educated son of George III's chaplain. Scott had no time for former missionaries, and persuaded Arthur that while the reverend 'did his duty to the utmost of such abilities as he possessed,' he was 'below mediocrity' and his 'age and infirmities rendered him quite unfit of so large and populous a district'.

Subsequent outrage in Launceston led the *Colonial Times* to report it had been 'requested to state' it was untrue Rev Youl's mooted removal caused him to die of a broken heart: medical officers had judged he died of 'a morbid inflammation of the bowels in consequence of indigestion brought on by eating too freely of peaches'.

Death by peaches! A man who gave his life to pioneer a school and church branded 'unfit' and 'mediocre'? Youl would never forget or forgive. It seeded a view that, stripped of uniforms and robes, all men were just men - and reinforced the Birkbeck edict to be beholden to no one. While his mother had clung to hope of him becoming a preacher, Youl had to 'save' the family, especially after his mother's bid to obtain a pension was rejected because it might set a costly precedent, and she declined the offer of a return passage to England.

This meant turning to the altar of money. Rev Youl had steadily acquired property but lacked the time and cash to properly harness it, and erect fencing to contain stock. He left some 4000 acres (1600 hectares), built up from a 700-acre (280-hectare) land grant on the South Esk as an establishment minister, compensation for the cost of his family's transfer to Van Diemen's Land, and offsets to foregone salary increases, plus 500 acres granted for James and his brother. On his death, the property ran about 500 sheep and 100 cattle, but it had potential for more.

Youl's neighbours understood potential. Henry Reed, the young son of a Doncaster postmaster, had just walked from Hobart to Launceston in pursuit of what he called 'decided success in spiritual and temporal things', which he quickly achieved. Reed might have counselled Youl that temporal wealth was part of service to God, with free land grants and convict labour a unique providence. 'Godliness is profitable unto all things', the Bible said, and men like he and young Scot David Gibson were successful followers. The latter arrived in 1804 as a convict sentenced to life for theft, but within 15 years of being pardoned in 1813 he held 7300 acres (3000 hectares) and entertained the visiting colonial governor Lachlan Macquarie.

Opportunities simply not available at 'home' allowed the creation of pastoral wealth and some of the finest Georgian and Italianate estates in the British Empire, with English names like Clarendon, Woolmers, Panshanger, Cheshunt, and Brickendon. A Protestant doctor freely advised 'get money, honestly if you can, but get money', and many ambitious settlers admitted 'we all came here to make money and make money we will by hook or by crook', then display their 'love of (mother) country' by returning with their wealth. An English woman visiting the colony decried the 'voracious spirit of money-getting', but for Youl temporal wealth offered the chance to 'save' the family in a 'Godly' way and perhaps eventually reward himself by returning to London.

First, he had to survive. Symmons Plains was part of ancient Aboriginal hunting grounds being usurped as pastoral runs in the most intense and deadly frontier conflict in Australia, with neighbour John Batman, the famed future 'founder' of Melbourne, leading midnight hunting parties. Youl protested when the government considered bringing back 'natives' forcibly removed to Flinders Island so some 'may be induced to submit to the restraints of civilisation'. He warned their love of 'unrestricted freedom' meant nothing would prevent a resumption of 'plunder', 'retaliations' and 'permanent hostility'.

Escaped convicts, including the infamous Martin Cash and Matthew Brady gangs and cannibalistic Thomas Jeffries, also roamed the sparsely populated and poorly policed plains. Youl helped form an association 'for the suppression of felonies', with substantial rewards for the serious crime of stock theft, for which many were hanged.

A year after his father's passing, four armed bushrangers confronted the family. According to Youl: 'A pistol was cocked and placed within a few inches of my head … my life threatened for something I had done which they did not approve.' An £80 reward and manhunt saw them captured in a shoot-out, then join 14 others condemned to death on a single day.

Youl sat on juries which often took just minutes, or didn't even bother retiring, to condemn men to death 'in the usual form'. Nearby Perth was where, for the final time in a British colony, a condemned man's body was gibbetted and left to rot on a tree as a roadside advertisement warning others to stay on the right path.

There was no room for sentiment in the pursuit of advantage. Youl sold his father's personal effects, including 150 books used in pursuit of 'intellectual and moral character' in the colony. For additional income or influence, he became postmaster, divisional constable, justice of the peace, and quarter sessions magistrate. Pound-keeping was an honourable post but allowed daily fees of up to two shillings a head for poundage, food and water. Accumulated fees soon exceeded livestock values, leading to protests that the pound system was nothing more than plunder and theft under the cloak of law.

In this Vandemonian 'stock exchange', Youl astutely traded stock, land grants, purchases, and leases. Others seeking access to land, using future crops as security, risked ruin: one man with a family of seven saw his animals, equipment and household furniture seized by Youl in an 'under distraint for rent' sale.

Youl was also astute enough to select from neighbouring families some prime merino offspring of royal flocks in Kew, known as the 'Hampton Court flock', and import superior pure rams from Germany. Selective merino breeding allowed Youl and neighbouring Cox and Gibson families to ride the sheep's backs as the island's sheep population grew from the first landing of 200 sheep in 1806 to a million by 1838. They laid the foundations of a golden fleece, Australia's famous Scone stud reference flock, and funded Youl's workforce of about 30 shepherds, gardeners, stockmen, agriculture workers, cooks, waiters and servants.

In a diary, a visiting Annie Maria Baxter, who as Annie Dawbin later authored *Memories of the Past by a Lady in Australia*, sniffed at the Youl 'purse pride', but it allowed him to build an impressive riverbank mansion. An avenue of elms, oaks and roses led to a 16-room, two-storey Georgian building with an extensive lawn and numerous outbuildings for coaches, carriages, horses, wool and grain.

James Youl's Symmons Plains home. Photograph, courtesy Cumulus Studio, Launceston.

It was an impressive home for the 27-year-old rising esquire and his bride, 21-year-old Eliza Cox, after their marriage on the opposite side of the South Esk at her brother James Cox's estate of Clarendon – perhaps the grandest in Australia, named after a palace in Wiltshire. Neither mansion would have looked out of place in the British countryside.

Nowhere in the world were people so far from their birthplace, and nowhere did distance make an English heart grow fonder for anything from 'home'. On a strikingly beautiful island, many eyes nevertheless saw only a landscape 'barren' of every species familiar to them. From the first convict ship in 1804, arriving vessels were something of an ark. In came sheep, cattle, horses, dogs, pigs, and poultry; oak, fir, poplar, elm, ash, birch, poplar; apple, lemon and pear trees ; strawberries, raspberries, cherries, potatoes, cabbages, lettuces, onions and peas; roses, carnations, and dahlias; pheasants and deer; swans and bees. The 'anything and everything' approach also landed the fox, rabbit, hare, starling, blackbird, willow, hawthorn, scotch thistle, blackberry, stinging nettle and gorse – and collateral pests like snails, slugs, wasps, moths and butterflies.

The insatiable demand was fuelled by isolation, insecurity, nostalgia, loyalty and ego. It was conspicuous to visitors: 'thoroughly English people clinging with tenacious affection to the memories and associations of home' (colonial judge Henry Francis); 'most delightfully resembles England' (Charles Darwin); 'junior England' (Mark Twain); 'everything ... is more English than England' (Anthony Trollope).

New maps defined counties called Cornwall, Buckingham, Devon, Dorset, Kent, Glamorgan, and Somerset. Towns and villages were christened Launceston, Ulverstone, Perth, Ross, Brighton Margate, Swansea, Perth, Ross, Devonport, Campbell Town, Oatlands, Melton Mowbray, Bridport, and Orford.

Youl's estate, in the parish of Chichester, in the county of Somerset, played host to cricket matches, included a pigeon loft housing 60 birds, and featured a peacock which often sat on the house parapet. He imported English flowers as a pioneer of the Launceston Horticultural Society, organised quadrille balls (with a 70-30 ratio of men and women in the colony, a gentleman should 'bring as many ladies as he pleases') and called esquires to the Bald Faced Stag Inn in Cleveland for the season's first hunt of a transplanted deer, or perhaps the pursuit of kangaroo held captive on the Clarendon estate of his wife's family.

Over port or claret, a gentleman could spend a pleasant evening talking about the day's cricket or tally-ho. But not emulate what they were reminded of in 1838 when a Launceston publisher, Henry Dowling, audaciously pirated Charles Dickens' first novel, *The Posthumous Papers of the Pickwick Club*, into a 25-part series. In addition to introducing his first convict character, Dickens had a Mr Snodgrass assure himself his unsteady condition was 'not the wine, it was the salmon'.

Ah, yes, the *salmon*. Just as the heavens needed stars, some insisted, a river was not a river without 'English fish', while others asked, 'what is a stream without its trout?' The rivers and streams of Van Diemen's Land – unlike mainland rivers, invariably flooded, muddy or dry – looked perfect for fly-fishing. Presenting an English, Scottish or Irish familiarity, they were named Esk, Derwent, Tamar, Liffey, Forth, Leven, Ouse, Mersey, Ouse, Don, Cam, Clyde, Dee and Shannon. But in His fine handiwork God had made a distressing error, forgetting to put salmon or trout in them. Native fish like grayling and *Galaxia* species were not English fish, and as British author James Bertram noted, 'there are men

Counties or land districts of early Tasmania, as named in the nineteenth century.

who never lift a rod except to kill a salmon ... great anglers will not waste time on any fish less noble.'

The southern land and sky could be easily 'furnished' with transplanted animals, plants, trees and birds, but rivers were a different story. A gentleman could not relive the days of setting forth into nature with rod and line, the excitement of an introductory tug, a hooked quarry speeding off like an arrow, darting and pausing, diving and leaping in a desperate bid to escape before finally giving up its good fight, and laying, at the feet of the victor, shining brighter than diamonds ... then enjoy a feast of what angler-novelist Charles Kingsley described as the finest 'of all heaven's gifts of food', with subtle, pink flesh flavours melting in the mouth.

Waters looked salmon-friendly, but there was no salmon. Early arrivals to Australia thought they spotted salmon, and christened Australian, colonial or blackback salmon, with the young known as cocky salmon, but salmon they were not – rather *Arripis* specimens with a superficial resemblance. But there were fly-fishers: in the first Australian reference to fly-fishing, in 1833, Welsh-born editor-surgeon Dr Thomas Richards described in *Hobart Town Magazine* a day's fishing in the Plenty, north-west of Hobart, where armed with a 'red hackle' he enjoyed a splendid day of 'murderous sport, which we pursued with unrelenting vigour, till our baskets [would] positively contain no more'. The grayling offered some sport and a pleasant meal, but there was fish and there was salmon.

The first public airing of the salmon dream came in the same year. After delivering the first hive of English black honeybees, Royal Navy convict surgeon-superintendent and Royal Geographical Society member, Dr Thomas Wilson, further excited the colony with a promise to bring out 'English fish'. The *Hobart Town Chronicle* said if the 'noble attempt' for salmon and trout was successful it 'will indeed commemorate his name to the latest ages of a grateful posterity'.

Wilson's promise faded, but not the dream, and it was re-ignited when Vandemonians began seeing brief but extraordinary newspaper reports in the 1840s about 'artificial production of fish'. Two poor French fishermen had seemingly found a way to have 'an inexhaustible supply' of trout, then a wealthy Lancashire manufacturer in Ireland produced 'no less than twenty thousand salmon fry' costing, he said, no more than a farthing each.

For a colony built on convictism, the Godly grace of angling and exquisite dining – long the exclusive province of priests, aristocrats, and the wealthy – and perhaps the prospect of a salmon industry, teased as a gift that could change everything.

For a sheep farmer like Youl, familiar with the costs of acquiring, feeding, breeding and managing livestock, the reports were astonishing. But he was preoccupied with his pastoral interests, didn't have any passion for angling, and knew no more about salmon than what he had seen in a London fishmonger shop. But the story of his life, thanks to salmon, was about to change.

2

THE TIDE TURNS

Forty years after banning British ships and citizens form involvement in the slave trade in 1807, critics in London were agitating for an end to convictism. The assigned labour system providing cheap labour had been replaced by a less punitive pathway to freedom, but when Van Diemen's Land became Australia's sole remaining exile gaol, convict numbers surged.

This coincided with depression and unemployment following a gold rush exodus to Victoria. Declining prices for wool and reduced home funding fuelled concerns of a population of criminals and paupers. Island squires like Youl could see the economic tide turning. It was clear the island needed to stand on its own two feet, without manacles and chains, and settlers chafed at the island's exaggerated reputation as a 'horrid place of evil ... the worst days of Sodom and Gomorrah ... not as bad as the present days in Van Diemen's Land', as the House of Commons was told.

With an increasing struggle to find and keep honest workers, Youl helped form anti-transportation petitions to Queen Victoria, and took a lead role at an 1850 meeting at Launceston's Cornwall Hotel, described by the Hobart *Courier* as perhaps 'the greatest ever held in Van Diemen's Land'.

Describing himself as a 'plain-spoken man', Youl was cheered when he said the plain truth was that London had not met a promise to cease sending convicts to any colony averse to it. The Home Government simply had to 'get rid of the evil which is hurrying us all into ruin'. Governor Sir William Denison loyally resisted the anti-transportation push, but it spurred colonies to unify for the first time, in a rising 'The Australias Are One' sentiment, to end transportation.

As convictism was coming to an end, and with the island needing new industry, the prospect of salmon salvation emerged. Men in France, Scotland and England were not only unwrapping some of the natural history mystery of a fish – long admired but also misunderstood and mistreated – but opening windows to a new world.

In a remote village in North-East France, a poor fisherman, Joseph Remy, was frustrated by not catching enough trout. Hiding in a tree to find the explanation, he could see thousands of ova and hatchlings were at the mercy of predatory fish, birds and insects, or any rush of water or mud. To change the odds, he set about experimenting with the help of innkeeper Antoine Gehin. They mastered the art of manually milking and fertilising male milt and female eggs, then

Fig. 543. — Remy et Géhin au bord de la Bresse.

Joseph Remy and Antoine Gehin shocked the scientific world with the results of their manual milking of female spawn and male milt. Illustrations, Les Merveilles De La Science ou Description Populaire des Inventions Modernes by Louis Figuier (Furne, Jouvet, c. 1870). Look and Learn Historical Picture Archive.

spent several years developing a protective nursery, before settling on a circular perforated tin box, resembling a sieve, which they anchored in their La Bresse stream.

'I placed them in tin boxes pierced with a thousand holes and between grains of coarse sand who bottoms are well lined ... in the Bresse river in a fairly quiet place,' Remy described. Then came a joyous day – 'regardez! regardez! Le poisson!' – when the pair saw their first 'little ones'. The number of hatchlings rose each season from hundreds to 'a multitude', before, in 1843, Remy revealed the 'discovery' of how to hatch an immense quantity of trout eggs. He believed such 'happy results' were worthy of government attention, 'at a time when the rivers are almost devoid of fish'.

But Remy and Gehin were seen by many as demented, the notion of peasants being able to harness 'artificial fish' dismissed. That changed five years later when the most influential scientific body in the world, the Academy of Sciences in Paris, learned they had accumulated 'between five and six million trout, aged from one to three years, and the production of this year will increase that vast number by several hundred thousand'.

Scientists were embarrassed: a year before they had dismissed naturalist Jean Louis de Quatrefages' report of experiments showing how fish eggs could be protected and 'sown like wheat'. Also ignored was economist Baron de Riviere, who told the Societe Royale et Centrale d'Agriculture that an industry he christened 'pisciculture'[1] was waiting to be created as a new 'branch of rural economy'.

Cultivating land and domesticating animals allowed man to 'reign on earth supreme', but fish had largely escaped his dominion – although the Chinese had long been collecting eggs, and in 473 BCE China, Fan Li wrote the first known document on fish breeding.

Incredulous envoys from around the world rushed to Paris to verify for themselves what American writer William Fry rated 'one of the greatest discoveries of ancient or modern times ... like all other great discoveries, now that we have it this seems the simplest in the world. The most elementary of ideas ... are often those which are last taken hold

---

2. 1   from Latin piscis, for fish, and culture, modelled after agriculture.

of. It took 6000 years ... now we have the solution in so simple a form that even the children of farmers practise it as easily as they tend a flock of sheep.'

The discovery was, in fact, a rediscovery. In the 14th century, French monk Dom Pinchon discovered the art of manually fertilising trout eggs and hatching them in wooden boxes buried in the sand. Then in the 1760s, experiments by German soldier-scientist Stephan Jacobi led him to write 'it would be no hard matter to breed trout in a place where there never had been before'. And 'our method of multiplying them may be very useful' when it came to salmon.

The Royal Prussian Society of Sciences rejected Jacobi's notion as a 'false chimera' but reports of 'the fish maker' spread and were eventually translated into Latin, French, and belatedly English – but without exciting great interest. Although King George III, sometimes derided as 'the mad King' was sufficiently impressed by his fellow German that he awarded Jacobi a medal and life pension. Otherwise, as Fry later observed, 'so much occupied were the people of Europe in the art and science of cutting one another's throats' that it was lost sight of until Remy and Gehin's rediscovery coincided with a growing awareness of decimated salmon stocks.

Across the Channel, leading English inventor-angler Sir Humphry Davy wrote in 1828 of his conviction that the Jacobi method 'offers a very good mode of increasing' any varieties of the *Salmo* genus and hoped 'some enlightened country gentleman' would pursue the 'very interesting and untouched field'. But anything to do with salmon had been the source of fierce argument for centuries. The 1215 Magna Carta sought some protective 'rights' for salmon but landed gentry, industry owners and anglers continued to fight for self-interest over public interest. Disputes over the causes of declining stocks and possible solutions were further prolonged because of salmon identity wars.

A Scottish poet, James Hogg, known as the Ettrick Shepherd, became convinced in the 1830s that the small fish he caught while herding sheep around Border streams was simply the children of adult Atlantic salmon. To him, it was a 'parr', which grew into a silvery 'smolt' to migrate to sea and return in the majestic shape of a grey-blue adult 'grilse' to spawn.[2] Others maintained it was nonsense and gave local names to what they saw as different fish. As Youl would discover, 'brothers of the angle' might be the best of men but of stiffness in opinion.

In turn, anglers faced men calling themselves 'scientist'. In the popular new realm of natural history, the views of ordinary practical shepherds or fishermen were dismissed as 'flat blasphemy ... the mad theory ... of infidels and heretics' threatening the reputational 'dignity' of emerging science. Such dignity and primacy were felled by a Scottish gamekeeper. Years of observation and hatching experimentation in freezing winters led John Shaw, of Drumlanrig, to shock the Royal Society of Edinburgh in 1840 with his proof of the parr-smolt-grilse stages. While rivals conceded and anointed Shaw as 'the Columbus of the salmon world', others saw salmon in a new light. Scottish nobleman Sir Francis Mackenzie, of Gairloch, sought propagation advice from Shaw and an ally, rising Swiss naturalist-embryologist Professor Louis Agassiz who had also experimented with propagation.

Mackenzie was thrilled to produce from the River Ewe a good number of salmon fry about three-quarters of an inch (just under two centimetres) long. He promptly declared 'the breeding of salmon, or other fish in large quantities is completely easy and ... millions can be produced, protected from danger and turned out into the natural element at the proper age.' He and Agassiz, who would go on to natural history fame in America, went further: any quantity of salmon spawn, properly impregnated and managed, 'can ... be carried to other streams, however distant' as safely as if it were deposited by the parent fish.

The sensational proposition fuelled dreams in faraway Van Diemen's Land. Retired Scottish naval captain, Frederick Chalmers, of Bagdad in the Midlands, immediately sought two baskets of salmon fry from Mackenzie, but was advised they would not survive being carted 80 miles (128 kilometres) to Inverness, or a journey to London, let alone across the world. An easier plan, 'one that cannot miscarry', the Scots said,

2.2   Parr from the old English parren (to enclose), referring to the marks on the side of the fish, resembling the bars of a fence; 'smolt' from old English smeolt, for bright and shiny; grilse perhaps from the Welsh glas for blue.

would be to despatch impregnated roe in baskets inside a ship's water tank. They 'will require no more trouble till you land in Australia', and the basket contents could then be carried in a pail to any stream where 'ninety-nine of the one hundred grains of roe will become salmon'.

Chalmers' plan did not eventuate, but a young Scottish survey clerk in Hobart, James Burnett, was so taken with the promise of sowing distant waters with fish ova that he visited Invershin to seek advice from Andrew Young, a fishing manager who had just completed 'an account of the artificial incubation of the salmon from a long series of experiments and observations'.

Young didn't share Agassiz and Mackenzie's view about it being completely easy to transplant fish ova 'however distant'. He was 'extremely desirous' of fly-fishing being much more accessible to the public, but doing so on the other side of the world was something else altogether.

Attempts anywhere to transport ova or fry had always failed, he told Burnett, and 'will continue to do so'. The distance to Van Diemen's Land, and the equatorial heat on the journey, were insurmountable hurdles, and even if ova could be landed any surviving fry would have 'not a shadow of a chance of success' as they would lack the natural instinct of returning to the rivers where they were bred. From his studies of salmon 'from the cradle to the grave', the only answer to the 'vexing' question was to 'carry out adult salmon only'. They could endure 'more fatigue and hardships than the young ones', live in fresh or saltwater tanks, and be resilient enough to survive a long journey with temperatures ranging from 30 to 80 (0°C to 27°C). On arrival, they would face fewer enemies than at home, especially 'their greatest and most destructive enemy, the trout'. And, if placed in a river to spawn, their progeny 'are certain to return in spite of all the obstacles, because they are the rivers where they have been bred'.

Other than attempting adult salmon, Young declared, 'all other attempts must be looked upon as building castles in the air.'

Back in Hobart, Govenor Denison, an Eton-educated engineer, embraced the salmon castle. He enjoyed angling and understood a possible salmon fishery could address the emerging need for life after sheep and convicts. He was also patron of the Royal Society of

Van Diemen's Land, the first outside Britain, seeking to develop the 'character' of the island. He had his private secretary, Captain Charles Stanley, also an engineer, prepare a paper on 'the introduction of salmon from the rivers of Scotland'. Stanley supported Young's suggestion of transporting smolt in a large saltwater tank, to be placed in the Derwent River upstream from Hobart at Bridgewater, where fresh and salt water met, 'so as to induce them to ascend the river'.

Understanding London wanted the least spending on the penal colony, Denison proposed a 'simple plan' of fish and ships: to bring out salmon smolt, or perhaps fry, on a convict ship. The Secretary of State for War and the Colonies, Lord Henry Grey, liked the pitch for an experiment costing 'but a little and the result ... if successful, may prove of great benefit to the colony'.

But the plan fell through. Instead of a convict ship, Burnett quietly persuaded Young to press for a 'regular Scotch smack properly fitted up with wells', and the engagement of Scottish fishermen to accompany the salmon. Young told Lord Grey a convict ship 'when crossing the line would

Governor William Denison, 1859. Photograph, © Freeman Bros. Studio, Mitchell Library, State Library NSW

very likely kill the fish', while a smack would have 'little or no chance of failure'. But the mooted cost of £800 (over £69,400, A$145,000 today)[3] was 'an obstacle scarcely overcome' for Grey, and the risk of equatorial heat 'a conclusive objection'. Nine months after receiving Denison's proposal, he replied with 'regret' that no further steps would be taken.

Denison pursued the matter with a new Secretary of State, Lord Newcastle, who turned to Gottlieb Boccius, an enterprising inventor who had also mastered the art of artificial propagation and cited success in 'restoring' streams such as the Wandle, Colne, Derwent, Lea and Wye with 'millions of fish' for numerous dukes and earls. His brief from Newcastle was not to send young fish, as Young advocated, but salmon and trout spawn in tubs on the barque *Columbus*.

When the 50,000 salmon and trout ova perished long before arrival in Hobart, many were convinced the quest was an impossible dream and would remain thus. But Boccius remained confident of ultimate success, and it was hard to ignore the prospect of producing 20,000 salmon for no more than a farthing ... even to someone like Youl.

He was not one of the angling brotherhood borne of Izaak Walton's famous seventeenth-century fishing 'bible', *The Compleat Angler: Being a Discourse of Fish and Fishing Not Unworthy of the Perusal of Most Anglers*, which equated angling and Godliness, and anointed salmon as 'the king of fish'. But his memory of a majestic salmon on a fishmonger's slab in London meant he understood those who declared that anyone who had not seen salmon had not seen what a fish should be, and if rivers were not complete without salmon, then Van Diemen's Land was not complete. And an industry based on salmon costing so little could reshape the island.

The more Youl thought about it, the more his interest grew. With more reports of what newspapers called 'artificial live fish' enlivening Northern Hemisphere rivers just as 'we cover our fields with corn (and) ... multiply our flocks!', Youl sensed the possibility of becoming a 'salmon maker'; 'I thought what a grand thing it would be to introduce this noble fish into the Australian and New Zealand rivers and determined to use every effort in my power to accomplish it.' And while other colonists

---

2.3   The British pounds of the day have been used throughout this book, with discretionary indications of today's values through Bank of England inflation calculators, as at the time of writing.

held the same dream, it would be easier to accomplish from London, which where his head was turning.

He felt he had done as well as he could, and wanted to ensure his children were well educated. For better or worse, Van Diemen's Land would never be the same again, even though half a century of convict transportation finally ended in 1853, and the island became known as Tasmania. Launceston's 10,000 people had more facilities than most British provincial towns, but Youl was frustrated that more was not being done to raise 'the character and intelligence of colonial youth' by opening their minds to science and practical discoveries, as his had been. He wanted to live out his days in London, envisaging a comfortable life with other colonists in gentlemen's clubs, securing some useful directorships and helping press frustrations over mail communications, recognition of legal tender, and naval defences.

At the end of 1853, he leased much of his land holdings and sold almost everything: his flock of some of the finest rams and ewes in the country, farming equipment, and 'the whole of the excellent household furniture' of plate, china and glassware, and a Broadwood pianoforte. In early 1854, after a 30-year absence, he was ready to make his way back to a beloved Babylon with an expectant Eliza, five girls and three boys.

Salmon was also on his mind. More so after an editorial in the *Cornwall Chronicle* said bringing salmon or perhaps trout 'might not pecuniarily repay the experimentalist', but an enthusiastic man would 'find his trouble more than compensated in the chronological éclat, that Mr So and So was the first person who introduced salmon into the rivers of the Britain of Australia.'

The lure of salmon stirred. If it could possibly be done, he could be that Mr So and So, one of those men known forever for creating something which survived beyond himself.

It would take something miraculous to find a way to have salmon in waters where it had never existed, but if it could be done that would surely stand the Youl name to the good once and for all. And as he had learned at his father's Sunday School about a lad helping Jesus in a 'miracle of the five loaves and two fishes', sometimes miracles did happen.

3

THE PRICE OF FISH

The voyage to London on *Fingal* immediately drove home the challenge of shipping any fish across the world. The barque ran aground in the Tamar River, equipment failed in treacherous Bass's Straits[1], several sailors were washed overboard, and monster icebergs south of Cape Horn saw everyone 'thinking the end had come'.

Expectant Eliza was too ill to ever leave her cabin, and the moment the English coastline appeared the captain hailed a passing fishing smack so she could be lowered in a chair and travel upriver in the smaller vessel with the family to Plymouth.

After the delivery of baby Grace, Youl moved the family to London's largest and most sumptuous hotel – the Great Western in Paddington, just opened by Prince Albert – then into a mansion in an enclave of upscale villas and gardens in Clapham Park on the southside of the Thames. In a

---

3. 1   So named by explorer Matthews Flinders, it later became Bass Strait.

Clapham Park mansion, similar to James Youl's Waratah House.

street lined with horse chestnuts, hawthorns and laburnums, the house comprised eight bed and dressing rooms, three reception rooms, a large dining room, library, and domestic office, and accommodation for four servants – a cook, housemaid, nurse, and under-nurse – plus a coach house, stabling for three horses, and a fowl house. A large shady lawn and garden featured choice shrubs and trees and a carriage driveway.

Youl called it Waratah House, after a red-flowering species endemic to Tasmania. It was a long way from sleeping as a young boy on the floor in the Sydney hinterland. He had not followed his father's pastoral path, but he was well satisfied with the rewards of a more temporal flock.

He arrived at the midpoint of an extraordinary decade of industrial and species revolution, bookended by the Great Exhibition of the Works of Industry of All Nations in Hyde Park's Crystal Palace in 1851, and Charles Darwin's *On the Origin of the Species* in 1859. The spirit of the times was the pursuit of putting Nature to the service of man.

Prince Albert portrayed the Crystal Palace exhibition as part of a 'sacred mission': 'Man is approaching a more complete fulfilment of that great and sacred mission which he has to perform ... to conquer nature to his use; himself a divine instrument.' God wanted it so, as the Bible exhorted: 'Be fruitful and multiply and fill the earth and subdue it and

Isidore Geoffroy Saint-Hilaire, acclimatisation pioneer, preached Man's duty to make all of Providence available everything to everyone. Wikimedia Commons.

have dominion over the fish of the sea and over the birds of the heavens and over every living thing that moves on the earth.'

The 'one world, one obligation to God' ethos of the Great Exhibition coincided with a new 'religion' called acclimatisation. A French zoologist-naturalist, Isidore Geoffroy Saint-Hilaire, preached Man's duty to make all of Providence available to everyone. Using his zoo in Paris as a showpiece, Saint-Hilaire founded La Société Impériale Zoologique d'Acclimatation and sparked a world-wide pursuit of economic and cultural advantage through global interchanges of species.

Dominion over fish was to the fore. Europe was shrouded by an 'evil of insecurity', as the *Daily News* described, with rising populations causing fears of unrest and war. The newspaper hoped salmon propagation developments meant 'the first and greatest of human liabilities and social calamities might be avoided … it will be our own fault if anybody starves for generations to come.'

The influential *Quarterly Review* observed, 'everyone is interested about fishes … the political economist, the epicure, the merchant, the man of science, the angler, the poor, the rich.' Youl was also interested, although he had no idea if his 'salmonisation' of the Southern Hemisphere could ever be achieved, but he was seduced by a powerful lure, one even

Prince Albert urged mankind to conquer Nature to his use. Photograph, JJE Mayall, 1860. Wikimedia Commons.

stronger than that famously suggested a century before by Jonathan Swift in *Gulliver's Travels*: that the world might canonise as 'the greatest benefactor of mankind one who could deliver corn or grass to a land where it had not grown before'. Anyone who could be the first to plant fish in another hemisphere would surely belong to another realm of fame.

Youl did not know if it could be done, or how. He variously admitted to knowing no more about salmon 'than a wagon horse', or 'the glass in front of me'. He could only follow the advice of earlier lectures in London to be unafraid to confront and resolve challenges. He had followed this approach at Symmons Plains, learning from agricultural science literature and correspondence with practical Scottish farmers to better breed sheep and harvest crops. In long lectures at the Midlands Agricultural Association, he cited his own experiments with manure and ash to boost productivity, declaring it the 'sacred duty' of every farmer to not overuse land without replenishment, not so doing an 'evil ... crime against society'.

Overfishing without replenishment was the time-honoured crime against salmon. Centuries of protective laws with severe penalties, including death, for destructive fishing practices had not stopped what some called a 'massacre of the innocents'. The pursuit of 'progress' saw the mighty Thames, which had long had numerous fishing villages from Gravesend to Teddington catching salmon weighing up to 80 pounds (36 kilograms), become the 'noblest sewer'. It had been 30 years since it was last able to sustain a dainty dish to set before a king; a stark contrast to the days when salmon across the country was once so abundant and cheap that newspapers reported 'even the peasantry turned their noses up at it', Cromwell's army commandants were 'compell'd ... not to force or compel any servant or apprentice to feed upon salmon more than thrice a week', and farmers 'fed their pigs with the dainty little fish'.

Two giant influencers, Charles Dickens and Charles Darwin, could hear the salmon crying. Dickens' periodicals, Household Words and All the Year Round, said the 'silent effects' of the age had grown from day to day until 'they burst upon our view as stupendous fact ... the gradual extinction of the salmon. The cry of "salmon in danger!" is now resounding throughout the length and breadth of the land'. Unlike what had impressed Youl as a young student, not one gleaming salmon on the slabs of Billingsgate fish market was now locally caught. A historian, Dickens said, would perhaps record 'the inhabitants of the last century destroyed the salmon.'

Dickens' lament that 'ignorance more frequently begets confidence than does knowledge' was well and truly evident with salmon. Ignorance about salmon's lifecycle and over-confidence in Nature's supply meant the century-old advice of Swiss philosopher Jean-Jacques Rousseau, for man to beware of his hubris when it comes to the natural world – 'everything degenerates in the hands of man' – was missing in action.

Charles Darwin saw salmon and First Peoples as 'inevitable victims' of 'more dominant' species. Photograph, Francis Darwin's *The Life and Letters* of Charles Darwin, 1887.

The same weaknesses were evident in America where, 'destiny' and 'progress' saw salmon habitats become 'barren' due to excessive and indiscriminate fishing, dams and rural and urban development. For centuries, so-called 'savages' never interfered with spawning, and only took what they needed, but white man was, as pioneering conservationist Richard Nettle warned, 'the destroyer ... blind to his own interests ... determined to exterminate the whole of the salmon tribe'.

Fish fascination was evident in the Zoological Society of London gardens in Regent Park, where new plate glass boxes on iron pedestals allowed Youl and his children to witness a 'menagerie of live fishes', including perch, carp, grayling, tench, and an occasional trout. For those who had only ever seen fish in a net, or on a fishmonger's slab, this was a star attraction, and helped inspire a Victorian craze for home aquariums.

Salmon was absent because it was hard to manage and not well understood. A largely hidden and extraordinary lifecycle meant much of salmon's biography had hitherto been one with large doses of myth and romance. As Blackwood's *Edinburgh Magazine* put it, many aspects of the salmon remained 'shrouded in mystery ... a sealed book'.

Few understood that a female might shed 10,000 eggs, but perhaps 10 survived to become adults – the rest victims of infertility, sensitivity to changes in water flow and temperature, mud, algae, predatory birds and cannibalistic trout cousins. Any survivors had to then transform themselves to leave their freshwater nursery to live most of their life in salt water, growing muscle and oil before changing physiology again for a magical navigation return home to breed, utilising a mysterious 'chemical

Charles Dickens questioned the extent of interference with nature. Daguerrotype portrait Charles Dickens, 1852, Antoine Claudet. Public Domain.

memory' of smell, magnetic field and ocean currents across hundreds of miles. Only the fittest and most determined could leap their way home over weirs and falls, escape indiscriminate fishing hooks and nets, and still have the energy to engage in a last-gasp breeding contest. After a winning Master Salmon beat off rivals to release his white milt, or seminal fluid, over thousands of orange eggs laid by his favoured Madame Salmon in a shallow stream bed, salmon had given their all. For many, the sexual conquest ended in death.

None of this was known to Youl, which meant he could not know how long salmon would consume and bedevil him, with years of hardship, heartache, and derision as a 'lunatic'. No sense that his 'mad' idea of landing the fish on the far side of the world was maddeningly complex, with competing ideas and opinions of scientists, pseudo-scientists, anglers, inventors and philosophers.

Setting himself to the task, Youl hit the books in a reading room at the Jerusalem Coffee House, an epicentre of shipping and commercial news of the British Empire since James Cook's days, and at his Clapham Park home. 'I read all the books I could get relating to salmon and their natural history. I dreaded failure if I tried single-handed from want of information and skills.'

With an explosion of interest in natural history, declining salmon stocks, and the lure of sowing water like wheat, there was no shortage of material. A bibliography of known books on angling, fish and fish culture grew from 80 works in 1811 to more than 12,000 by 1883.

From Scotland, John Shaw's *Experimental Observations on the Development and Growth of Salmon Fry* was taken a step further in Sir Francis Mackenzie's *Brief and Practical Instructions for the Breeding of Salmon and other Fish Artificially*. Andrew Young followed up his *Growth of Grilse and Salmon* with *The Natural History and Habits of the Salmon* and co-wrote *The Book of the Salmon* with Edward Fitzgibbon. William Brown's reports of propagation experiments at Stormontfield became *Natural History of the Salmon* and Edinburgh anatomist Robert Knox produced *Fish and Fishing in the Lone Glens of Scotland: With a History of the Propagation and Growth and Metamorphoses of the Salmon.*

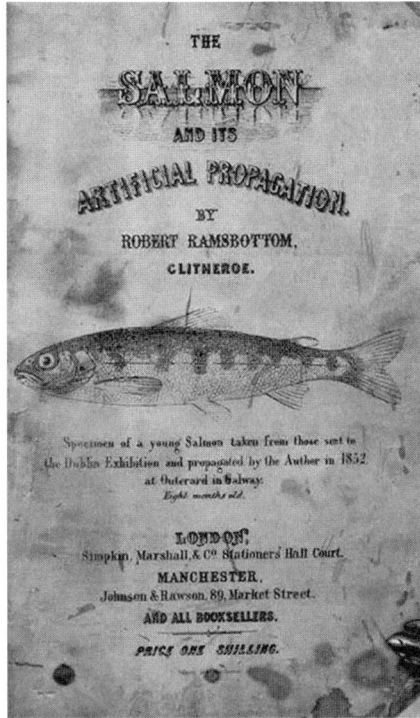

Robert Ramsbottom's *The Salmon and its Artificial Propagation*, was dismissed by some as reflecting views 'beyond belief by any sane naturalist'. Courtesy Oughterard Heritage Group, Galway.

In England, Gottlieb Boccius compiled *A Treatise on the Management of Fresh-water Fish*. The *Field* angling editor Francis Francis penned *Fish-Culture, a Practical Guide to the Modern System of Breeding and Rearing Fish*. From their fishery work in Ireland, Thomas and Edmund Ashworth published *A Treatise on the Propagation of Salmon and Other Fish* while their ova provider Robert Ramsbottom produced *The Salmon and Artificial Propagation*.

From France came translated reports of Joseph Remy and Antoine Gehin's techniques, Professor Jean Victor Coste's *Instructions Pratiques sur la Pisciculture*, Auguste Jourdier's *Pisciculture*, Haxo d'espinal's *Fecondation Artificielle et eclosion des Oefs de Poisson*, Jules Haime on *La Pisciculture* and Charles Millet on *Repeupler les eaux de la France*.

From North America, Dr Theodatus Garlick's experiments with trout in Ohio led to *A Treatise on the Artificial Propagation of Certain Kinds of Fish*, and William Fry compiled *Artificial Fish Breeding*.

Youl found himself drowning in information but thirsting for answers: could any salmon, be it spawn, young or grown fish, survive outside their home water and endure a voyage, longer they had ever undertaken through tropical heat they had never experienced, to live and breed, for the first time, in an alien hemisphere?

Each question spawned another. How could any fish be kept alive on a ship without natural water and food? Could freshwater fish survive three months in unnatural captivity? Could eggs endure two winters and a tropical summer within three or four months and survive temperatures between freezing and 95 (35°C)? How vulnerable were fish to variations in water and air temperature, oxygen, light? Could the new commodity of ice be safely used? Was it true, as some suggested, fish could survive being frozen? Would new hatchling, fry or adult fish be more viable? Should fish be transported in large tanks or small tubs? Made of wood, tin or iron? Could eggs or fish survive in still water or need a form of constantly fresh running water? Were rough seas and the tropics insurmountable obstacles? If any spawn hatched on board, could they be kept alive? How could any young or grown fish be managed and fed? If any fish or ova survived the voyage to Tasmania, could they survive in unfamiliar waters?

More broadly, despite propagation developments, critics maintained artificial fish breeding was just the 'mad theory' of 'feeble drivellers', a disparaging term for those so possessed of an idea they exhorted their 'conjecture and romance' to the annoyance of even close friends. Youl's bolder quest to transport salmon beyond the seas was beyond mad theory, more akin Jules Verne's science fiction. As one observer wrote, the prophecy of 'all the very wise people was unambiguous ... it could not be done; it was impossible to take the salmon to Australia'. If God intended salmon to be in the Southern Hemisphere he would have put them there.

The undeniable reality was that transporting any live fish over any distance, even on land, was a challenge. Taking saltwater fish on a 16,000-mile (25,700-kilometre) voyage was a bigger challenge. Taking less hardy freshwater fish was said to be 'a problem apparently beyond the power of human skill to solve'.

As the 'impossible' presented a serious financial and reputation risk for one acting alone, Youl cleverly secured support by calling former colonists from Tasmania, Victoria and New South Wales to the New Hummums Hotel in Covent Garden in March 1855. They formed a General Association for the Australian Colonies to press colonial issues, such as improved postal services and non-convict emigrants. But as secretary-treasurer Youl also secured backing of his salmon quest. He persuaded the association to embrace Tasmania, with its climate, rivers and lakes most akin to the Mother Country, as the first nursery of choice. If successful it would become a base station to dispatch the fish to other colonies of Australia and New Zealand, perhaps as 'the sole colonial emporium'.

The ambition drew derision. Youl heard he was 'insane', a 'lunatic' talking 'humbug'. He pressed on, but all the books did not take him far, as no one had done what he imagined. As science pioneer John Herschel put it, he needed a 'readiness to dismiss idle prejudices and pre-conceived notions' of both scientists and practical men. And Abraham Bartlett, a self-educated friend of Charles Darwin and natural history supervisor at the new Crystal Palace in Sydenham, stressed that 'however much we learn by books or words, it is unequal to that which we witness as a means to acquire knowledge'.

The only 'witness' was Gottlieb Boccius, the one man with any experience at attempting to transport salmon and trout to the Southern Hemisphere. Youl met the red-haired inventor several times to better understand his attempt with the *Columbus* and why it had failed.

Despite the 50,000 salmon and trout ova being dead on arrival, Youl was buoyed by Boccius' conviction that without issues beyond his control 'I have not the slightest doubt that this first experiment would have been perfectly successful .... I am now quite satisfied that salmon spawn can be conveyed in perfect safety alive'.

Boccius's biggest fear had been preserving life through the tropics. Celebrated natural history scientist Professor Richard Owen had suggested that 'if practicable' ice would be of service in a spawn tub. Commercial ice was a relatively new commodity, coming by the shipload from Boston Harbour, and was managed by a suitably named John Salmon. Queen Victoria quickly installed an ice cellar to chill and preserve her food, drinks and ice cream, but Boccius eschewed ice as a fatal risk. Instead, he placed ova in a nursery bed of gravel in a 60-gallon (227-litre) oval-shaped wooden tub, cased in lead, slung under the fore hatchway, giving *Columbus* Captain Daniel Smith orders to add six gallons (23 litres) of fresh water every six hours, then every four hours when the ship neared the equator.

He explained to Youl he had calculated 100 days in 54° (12°C) for incubation of salmon spawn, and 50 for trout, not anticipating any hatching until mid-April. But a delayed departure and contrary winds in the Downs saw Captain Smith report 'the premature appearance' of trout and salmon on 29 February, well before crossing the equator. Then, off the West Coast of Africa, the *Columbus* endured intensely hot weather for two weeks in 'the doldrums', the windless conditions caused by converging trade winds. Curious passengers watched newly hatched fry 'in a half torpid state' on the gravel bed and tub sides, and the water become more 'thick and putrid' as the heat took its inevitable toll.

Despite the failure, Boccius boasted he was the first to breed salmon and trout on a ship at sea, and had proved the fish did not require constant fresh running water, 'the theory insisted upon by philosophical writers and practical men'.

Youl was not convinced about any absence of running water, and the two disputed whether ice was a fatal attraction. Andrew Young had accidentally discovered ice could extend incubation when he found a severe Scottish frost merely delayed hatching of ova, but Boccius argued ice was a killer if the temperature reached freezing point. That was why, he said, fish 'instinctively choose the head or fall of a rapid ... its spawn being less likely to become frozen.'

While Boccius maintained that ice would 'kill the ova', Youl disagreed, later telling colleagues 'I came to the conclusion that ice must be used to cool the water in passing the tropics.' Whether Youl was independently of that view, it had been reached by others even before he left Tasmania. English botanist John Bidwell, in Australia as the mooted first director of

Sydney's Botanic Gardens and a friend of the Denison family, felt obliged in June 1852 to offer his advice: transporting grown salmon would never succeed, as other than carp or goldfish, 'no other exotic fish had been successfully established as colonists in any country'. He put this down to the small numbers involved: '(But) if thousands could be liberated at once, the chances would be in favour of any fish establishing itself in a new river in any suitable climate. To do this, it would be necessary to bring and hatch the spawn, and I think that by packing spawn in ice there would be no difficulty in preserving its vitality for a much longer time than would be required.' He did not see ice as fatal: 'By merely packing it in ice there would be no danger of actual freezing, as the ice would always be in a melting state.'

A month later, James Burnett's report to the Royal Society on the 1852 *Columbus* failure also advocated ice as an ally.

Youl may have been unaware of the Bidwell and Burnett views, which were then effectively lost for years within Royal Society files – but when they were fully heeded a decade later, the ice idea had many competing fathers.

For now, Youl was absorbing the conflicting opinions of scientists and fishery men about water and ice, and of spawn versus fry or grown fish, into a growing instinct: a lot of spawn could be transported in less space for less cost than hatched or older fish, and the more fish that could be landed, the greater the chances of success.

He just had to unlock the secret of 'how'.

4

THE LURE OF REWARDS

Of many questions to resolve if anyone was to ever turn the salmon world upside down, the first was money. It might be a noble mission, but much as he wanted to be the Mr So and So, Youl was not a man given to sacrificing hard-earned pounds for noblesse oblige.

He envisaged as much as £1500 (£178,000 today, A$372,000) would be required 'to make any attempt likely to be successful,' and he was not going to spend 'so large a sum myself for an experiment from which I would derive no benefit'.

He persuaded his Colonies Association colleagues that, as Tasmania stood to gain a considerable public benefit, it ought to be a substantial backer. To press the case they engaged James Gibson, chief agent for the Van Diemen's Land Company, a group of London merchants granted 250,000 acres (101,170 hectares) to supply wool to Britain's textile industry, to approach Dr Joseph Milligan, the company surgeon and, more importantly, secretary of the influential Royal Society in Hobart. There was confidence the Society, with the governor as patron, would favour funding the quest and support a committee – Youl and pastoral associate James Maclanachan, and VDL Company directors Gibson and Frederick du Croz – to 'carry out the matter'.

The cost of delivering 50,000 spawn was put at £500 (£59,500 today, A$123,000), much less than what Youl cited, 'but we think if you can remit us £450, that we shall be able to raise the other £50 here, ten subscribers having already put their names down for two guineas each'.

The Royal Society was taken aback. The request was half its projected 1857 expenditure, and wealthy individuals in London were committing a 'paltry' two guineas each (£190 today, A$400). And it was to support shipping of spawn when Andrew Young's adamant advice from Scotland was that only 'old fish' had any chance of surviving the arduous voyage.

Milligan turned to Colonial Secretary William Champ, a pioneering Society member and soon to be Tasmania's first Premier. Champ advised the legislature was prepared to pay a £500 reward to anyone who could introduce 'enough salmon or salmon spawn as would afford them a fair chance of being perpetuated'. And he pointed out that in New South Wales, Hunter River businessman John Reeve had also offered to match any reward of £500 or more by any colony 'in Australia, Van Diemen's Land or New Zealand' to ensure failure was not due 'the want of sufficient funds'.

With at least £1000 (£96,300 today, A$201,000) on offer, the Society told the London colonists 'it did seem scarcely necessary' to do more than suggest they spend £20 to publicise the reward. But even this would not be easily won. The Society said it ought to be paid only on delivery of at least 500 living salmon fry, at least three months after their hatching, or at least 250 salmon smolt 'alive and healthy' and ready for a journey to the sea, or five pairs of full-grown male and female salmon which had visited the salt water. The mere introduction of any spawn 'ought not of itself entitle the person introducing it to any portion of the reward'. It should only be paid on the first proof of salmon 'having actually spawned and bred in any Tasmanian river'.

And while the Society said it did not want to get into the minutiae of arrangements, it called for the shipping of spawn, young salmon fry, and salmon smolts. It also wanted the government to liaise with Burgh Magistrates in Perth to find 'a thoroughly experienced and trustworthy person' and one or two 'active, intelligent and well-behaved men of experience' for the voyage and run a pisciculture station for three years to fully test the practicability of distributing salmon into the island's rivers.

Further, notwithstanding the long yearning for 'English fish,' the society suggested the government consider the view of sea captains that salmon from California and British Columbia offered a greater chance of success and save the 'very considerable' cost of a longer voyage from England.

Not for the first time, Youl was irked by the Hobart elite. He chafed at the prospect of his quest being determined by 'rules' written in Hobart, where the government in early 1858 asked Royal Society fellows, including Dr Milligan and a lawyer-natural history devotee, Morton Allport, to 'frame rules or instructions touching the management of salmon spawn and salmon fry from England to the colony'. Furthermore, a reward was not committed funding, and, as Youl saw it, he and his fellow colonists freely offering their energies on salmon and other fronts were seemingly deemed not up to the task. And North American fish was not English fish.

With the colony so expectant of a salmon treasure that it had already drawn up salmon protection laws, Youl was annoyed that expectation did not come with expenditure. A £500 reward was 'altogether inadequate to remunerate or induce anyone to go to the expense necessary to undertake the experiment with any chance of success'. He told Dr Millligan a much more generous reward was £500 and at least 5000 acres (2000 hectares) of land. And using words that could easily have come from Youl himself, an anonymous contributor to *Field*, 'the country gentleman's newspaper', said anything less than £5000 'would not be too much for the trouble, risk and anxiety of the speculation, to say nothing of money sunk in machinery, outfit and passage etc.'

Nevertheless, the reward raised eyebrows around Britain. 'One hundred pounds for a pair of salmon!' exclaimed *Field*. It was seen as a tempting fortune, one that could induce many in England and France to make the attempt. Some in Tasmania saw a £500 reward as stingy, and altogether 'selfish and unjust to the last degree' to expect a few colonists in London to carry the whole responsibility and cost of transplanting any species to the colonies, but it was argued the colony's poor financial state meant it was 'rather too much to risk on an experiment so precarious'.

Until the public-private money question was resolved, Youl continued salmon plotting while dealing with other matters. His wife delivered another baby, Octavia, but the child died after fifteen weeks. And despite frustrations with Hobart, he felt he could not reject a plea from an old cricket friend, William Henty, who had become Colonial Secretary and could perhaps deliver more funding. Youl agreed to organise emigration promotional displays at Crystal Palace in a bid to combat the convict reputation stain and arrange what would be seven shipments of 'first-class' female domestic workers.

Youl also had a busy Colonies Association agenda, seeking British support for emigration, a 'federal assembly' to handle intercolonial matters, representation in the imperial legislature, and surety about military protection. At celebrations of the 26 January 'foundation of the Australian colonies,' he joined unending toasts to the 'Mother Country' and proclamations that the race of England 'must be preserved pure in the Australian colonies'.

Alongside salmon, his most public campaigning was to speed up mail to and from Australian and New Zealand colonies by using steam vessels and a Suez overland short cut instead of ships sailing around the Cape of Good Hope. Interest in both topics was so intense that colonial newspapers carried updates under 'steam and salmon' headlines.

Whatever the funding outcome, Youl weighed the advice of experienced Scots that the mission was problematic enough even with 'the most trustworthy and skilful hands'. They were deeply pessimistic, but supported at least making an Antipodean attempt, declaring that 'he who is the means of stocking the rivers of Tasmania and New Zealand with salmon would deserve his country's thanks.'

Even before leaving for London, Youl had noted the name of Robert Ramsbottom as a likely accomplice, reading reports of his salmon propagation work in Scotland, Ireland and England. Ramsbottom was from Lancashire's Ribble Valley, where isolated monks once found security from 'the intrusions of the world' amid green slopes and the bounding Ribble with fresh and living waters teeming with the rich and silvery trout and salmon.

When the region became the cradle of the industrial revolution, with Manchester the Cottonopolis of the World, Ramsbottom and his son William escaped the manufacturing roar by trekking to a peaceful stream in the Ribble to pursue 10-pound salmon and trout. Drawing inspiration from seeing raw cotton or calico transformed through design and colour into a desired object, Ramsbottom began using coloured scraps, dye, feathers, and hair to create eye-catching artificial flies. His art of deceit did well enough for him to become a 'fishing tackle manufacturer' in Clitheroe.

Some of his wealthy customers had bigger ideas. Cotton king Thomas Garnett, from nearby Low Moor, had been one of the first, in the 1830s, to air the possibility of artificial propagation in London's *Natural History Journal* and *Gardener's Chronicle*. He placed salmon fry into an artificial lake of Yorkshire friend Francis Fawkes of Farnley Hall, near Otley, where artist JMW Turner was a frequent angling guest. When the fry grew to a good size and afforded excellent sport, Fawkes quickly wanted more, as did other esquires.

When Garnett dangled propagation opportunities to the man he described as 'one of the best anglers and fly-makers between Cornwall and Caithness', Ramsbottom was willing to bite. Afterall, Napoleon had richly rewarded Joseph Remy with a state-licenced tobacco shop and an annual life pension of 1200 francs (estimate £4800, A$10,000), and tasked Antoine Gehin with the responsibility of 'vulgarising and popularising their process' with a travel allowance, reward of 1200 francs, a tobacco shop and a 500-franc (estimate £2000, A$4000) annual pension.

Fish-making looked more rewarding than fly-making, but the Ramsbottoms found that those who said manual propagation was 'as easy as shelling peas' had not shelled peas from slippery fish in the freezing water of a spawning season. Layers of clothing and trouser leathers could not stop the numbing of hands and fingers as they learned how to best obtain and fertilise spawn and develop protective spawning boxes.

Even finding and netting a suitable female and male fish, healthy and ready to spawn, was not easy. Sometimes it meant netting under moonlight at 3am. William held the tail of a frantic female fish while his father held its head steady and massaged thousands of her pale orange eggs into a small tub or tin of river water. Any white coloured ova, indicating sterility, had to be delicately removed with a fine pincer so they did not spoil the rest. Then a male's milt was milked onto the eggs and the contents of the tub hand stirred. After a minute the water was poured off, save enough to keep the ova submerged, and fresh water added. This was repeated several times.

The eggs then became brown-ish in colour with a tiny black spot in the middle, indicating fecundation was complete. The spawn was then carefully placed in a nursery the Ramsbottoms devised, similar to that of Remy and Gehin: a small box, made of wood or zinc to avoid rusting, perforated with holes at each end so water could run continuously through it. Spawn was gently laid onto a bed of small gravel and covered with some pebbles, and the box secured.

Several weeks of meticulous midwifery followed, using a bird feather to separate any eggs sticking together, a brush of badger hair to clean away any sediment, and pincers to remove any dead eggs. Then came a finny miracle: the gradual emergence of two tiny black dots, the beginning of eyes, then the faint sign of a tail emerging, the hint of lower fins, then a head and a body spinning inside the egg. After hatching, a yellow umbilical pouch under their belly provided sustenance. When

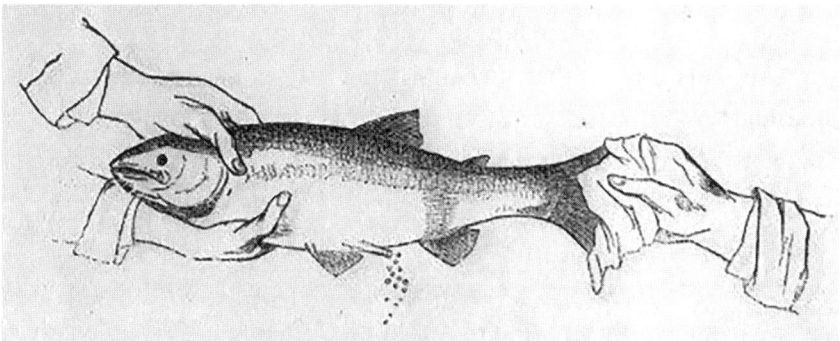

Milking spawn by hand.

this expired, the small fry were kept in the box for protection and fed frog spawn or shredded intestines of sheep and oxen until they were ready to be released into the river.

After mastering the 'art' of propagation, the Ramsbottoms filled Garnett's artificial pond beside the Hodder with 5000 salmon. Garnett declared it 'the first successful attempt with salmon ever in England'. Other cotton men could see golden ponds. Quaker brothers Edmund and Thomas Ashworth, of Bolton, wanted to 'render the science of pisciculture subservient to commercial purposes', buying a neglected fishery, Oughterard, on the Corrib in Galway. They engaged the Ramsbottoms to reshape stream and pond water to lay down 40,000 eggs transported from Scotland – the largest number in a single operation in Britain.

At the Dublin Exhibition in 1853, father and son proudly watched Queen Victoria and Prince Albert's fascination with a small glass tank containing some of their 'small but beautiful' salmon fry. The royal couple frequently spent their summer vacations fishing in the River Dee at Balmoral, and Albert immediately requested reports on fish propagation and preservation.

Scots were embarrassed their salmon 'home' had been bested by the French triomphe with 'artificial live salmon' and Scottish ova seeding the first major commercial propagation in Ireland. It led to a plan for replenishing the River Tay with Salmon and a Thomas Ashworth briefing on the Galway success. He said it was 'as easy to propagate salmon as it is to raise silkworm on mulberry trees', and as profitable to breed salmon as sheep, his salmon costing not more than a farthing for each. He pressed the importance of sending for Robert Ramsbottom.

The Scots did just that, and after Ramsbottom hatched hundreds of thousands in Tay ponds at Stormontfield, the Perthshire *Courier* said: 'It is now beyond a doubt that a fishless river can be stocked artificially with any kind of fish.' Others were excited: 'the eggs of salmon may be as carefully hatched as those of fowls.'

Ramsbottom was rewarded with 20 sovereigns and the Ashworths making it known that to 'Ramsbottom of Clitheroe ... first to conduct the proceedings in each of the three divisions of the United Kingdom

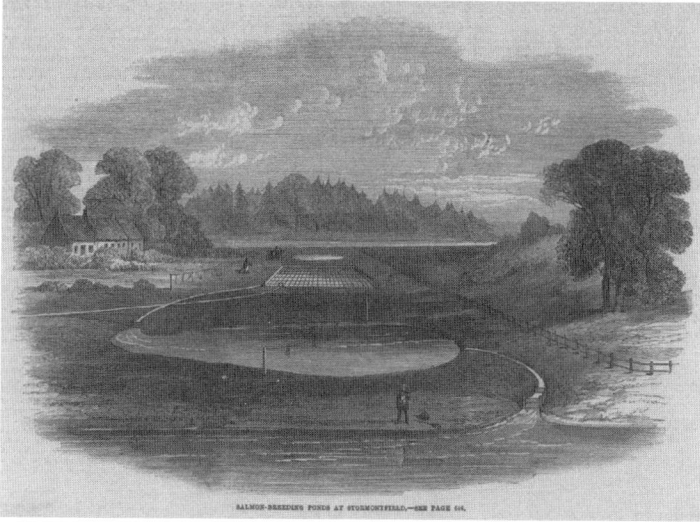

Salmon breeding ponds at Stormfront, Scotland, 13 June 1863. *Illustrated London News*. Courtesy Look and Learn Historical Archive.

... (and) we are indebted for the practical development of this science.'

Ramsbottom's subsequent small book, *The Salmon and its Artificial Propagation*, coincided with Youl's arrival in London. Bell's *Life in London* dismissed the book as the product of one of those who 'must be looked upon as dreamers', with views of salmon breeding 'beyond belief by any sane naturalist'. Others said men like Ramsbottom had 'libelled' the salmon as the only one of God's species incapable of fulfilling its own existence. Using nom de plumes such as Salmo and Piscator, they maintained artificial propagation was unnecessary if salmon was adequately protected by fishing laws 'for the first time in the memory of man'. One contributor to *Field* said, 'it's no' natural to middle wi' them, you should never middle wi' nature.'

Youl, familiar with ridicule for his 'act of folly', was encouraged by the admission of influential Tay superintendent Robert Buist that he had 'a good deal of misgivings as to (the) practicability, utility and beneficial results' of propagation before Ramsbottom's 'great knowledge' had shown the practicability of artificial breeding of salmon 'without a doubt'. The director of France's new breeding facility at Huningue,

Professor Victor Coste, also made it known that the leading men to talk to in England were Ramsbottom and those he had propagated for.

Buist said the 'darkness that prevailed on the natural history of the salmon' had been lifted and it was 'astonishing that the fish had been under our eyes for hundreds of years as a source of national wealth'. Not only could man 'successfully cultivate the waters as he cultivates the land', but to leave everything to nature, as some argued, 'is just about as reasonable as to say that we ought to leave our fields to sow themselves'.

Buist was quickly inundated with 'how do we get the salmon?' inquiries from across Europe and distant colonies of Australia, New Zealand and America, and the 'art' of propagation gave way to 'science', and the language of the land – 'seeding', 'breeding', 'farming', 'cultivation' and 'harvesting' – transferred to water in a new branch of 'rural economy'. And what an economy, as the Ashworths declared: a hundred salmon producing a million eggs 'which, by protection from their natural enemies, become fish'.

Such language and economics appealed to Youl, and he also warmed to Ramsbottom's views on his Antipodean quest. After inspecting some Norwegian propagation stations, Ramsbottom wrote to *Field*: 'I have no hesitation in saying that it is quite possible to transport both marine and freshwater fish to Australia.' And given the public benefit, 'our government should be lending a helping hand.'

Ramsbottom also seemed to have a like-doggedness. If his own obsession – to 'save' salmon with what he called 'the final cure' of new laws to combat poaching, unseasonal catches and fatal net mesh sizes – led to him being described as a 'discoverer', it would only be as foremost preacher Rev Sydney Smith put it: 'Man is not the discoverer of any art who first says the thing; but he who says it so long, and so loud, and so clearly, that he compels mankind to hear him – the man who is so deeply impressed with the importance of the discovery that he will take no denial, but at the risk of fortune and fame, pushes through all opposition, and is determined that what he thinks he has discovered shall not perish for want of a fair trial.'

Youl, 'deeply impressed' with the importance of his own quest, engaged Ramsbottom as his spawn man, describing him as 'one of the most efficient practical propagators of salmon, who has been engaged for thirty years in all parts of Europe.'

What he did not know was that while Ramsbottom proclaimed freshwater fish could be sent across the world, he did not believe in transporting salmon ova. In his mind, 'you might as well try to fetch Australia to England.'

5

A SALMON RUSH

'Artificial' fish seemed to deliver the biblical promise that 'the riches of the sea will come to you'. As Thomas Garnett put it: 'We send a little fish down to the sea, which is not worth a penny, and he remains there paying neither rent nor taxes, nor gamekeeper nor bailiff's wages, costing nothing to anyone until he returns to the river worth ten or twenty shillings. Surely this is a branch of the public wealth that deserves sedulous cultivation.'

While Tay superintendent Robert Buist, who liaised with Charles Darwin on the life of salmon, was convinced 'the seed of salmon (can be) deposited in boxes like peas in a garden', the emerging pisciculture science, he said, still required 'men of genius' to provide the general population with the wholesome diet enjoyed as an 'exclusive luxury to the rich'. But he saw the possibilities, just as 'the fall of an apple from a tree and the raising of a tea-kettle lid by the steam of boiling water gave hints which ... eventually produced the mighty discoveries of a Newton and Watt.'

Youl's bold ambition appealed to Buist, but he struggled to see a salmon garden on the other side of the world. He was certain the only way salmon could possibly be introduced into distant colonies was by getting ova to its destination before it hatched, but saw this as an 'almost hopeless' challenge. Delicate young fish would quickly come to life in warm weather and 'cannot possibly be saved unless a stream of fresh water be kept constantly flowing over them. This could hardly be done aboard a ship.' Any hatchlings 'could not be kept alive' to survive such a distance, even if near-freezing water was available.

Youl was not lacking people telling him his venture was a hopeless impossibility. But if anyone insisted on persisting, the Scots offered some suggestions. If there was to be any chance, first and foremost, 'the ova must be kept, by ice or otherwise, at such a temperature as the average cold of our Scottish winter.' Water had to be fresh and pure ('the purer the better'), well aerated, and the temperature kept under 45°F (7°C) at all times. 'To attain this degree of cold, while in the tropics, ice will require to be used.'

As for a possible transport apparatus, William Brown, an angler and secretary of the Perth Literary and Antiquarian Society was tasked by Stormontfield to help field correspondence and compile a history of its experiments, provided a sketch and description. Water, controlled by a stopcock, would descend from a cistern or cask through pipes and filters to be aerated before being deposited into a hatching box. A layer of ova, deposited between a base of sand and a covering of gravel, would be cooled by pieces

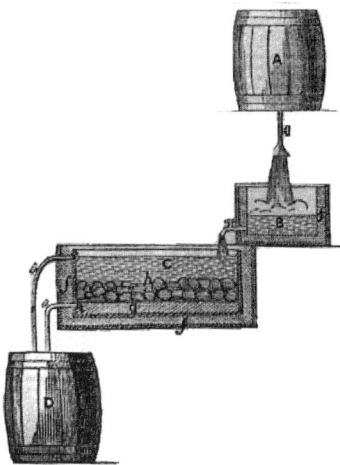

Apparatus suggested by the Scots to help ship salmon to Australia, 26 December 1857. Illustration, *Field* Magazine.

of ice in the box, and if necessary, in the supply cistern. To keep the ova stable and secure while a ship was rocking and rolling in rough seas, the Scots suggested an apparatus fixed to the framework 'suspended by elastic cords'.

The Scots' calculation was that in 45°F (7°C) water, the ova would take 60–80 days to hatch, and for six weeks afterwards young fry would require no feeding beyond their embryo sac – and 'by that time the passage to any of those colonies could be made'.

But as with anything about salmon, everything was passionately and forensically contested: the merits of ova, fry or smolts; matters of water, ice, oxygen, aeration, sunlight, piping, gravel and the nursery apparatus.

To help his thinking, Youl turned to the new Crystal Palace at Sydenham, the 'Palace of the People' offering the public levels of entertainment and education previously only within reach of the wealthy. Amid landscaped gardens, statues and fountains was a zoo, bird and agricultural exhibitions, pantomimes, circuses and concerts. Youl had helped the palace's natural history supervisor and taxidermist, Abraham Bartlett, with Tasmanian exhibits in the popular Australian collection of kangaroo, platypus, possum, emu, and models of two 'natives' portrayed as 'savages ... a degraded form of man'.

Youl wanted experiments to explore the impact of water temperature on ova survival, knowing he had to shepherd ova through the tropics. Bartlett was also urged by Rev David Esdaile, a keen natural history writer whose brother helped instigate the Stormontfield trial, to add a salmon exhibit following the enormous popularity of fish ascending ladders at the Dublin Exhibition and

Abraham Bartlett believed leaping salmon at Crystal Palace could be as big a drawcard as his famed gorilla. Photograph, 1858. Wikimedia Commons.

the Exposition Universelle in Paris. Encouraged by American circus pioneer PT Barnum, with whom he traded species from around the world, Bartlett could see leaping salmon being as popular as his 'newly discovered' gorilla from West Africa. He developed a small 'model pond' and 'river' for boxes for Tay ova, which arrived with firm instructions to keep the ova at no more than 40ºF (4ºC).

In January 1858, *Field* described an 'ingenious and interesting project' but questioned what knowledge of salmon possessed someone to commit the 'grave' error of putting the trial in the tropical department. If by some miracle any ova were to hatch, *Field* declared, hatching salmon would be 'on a par, if it be not somewhat easier, than that of hatching chickens, and any bungler, any old woman who has a cistern of water at her command will, without difficulty, be able to hatch her own salmon.'

There was no miracle. The ova perished. *Field* judged the 'bungle quite unworthy (of) even the name of experiment'. Perthshire Scots who supplied the ova were outraged, declaring palace 'naturalists' were just 'naturals', and their pessimism about Youl's quest deepened: 'We are not of the opinion that the ova or the fry of salmon could be transported with chance of success.' Ova was too delicate, and while well-attended fry might survive the journey, 'the care required (would) be so great the result of success we think hopeless'.

Youl still favoured sending ova, and had a suitable collector in Robert Ramsbottom ready, but not knowing how to get salmon all the way to Australia was giving him doubt and despair.

It was also aggravating that a rival Mr So and So of salmon was active in Hobart: Morton Allport, a successful lawyer, amateur naturalist and influential Royal Society member. As early as 1842 his father, Joseph, constructed ponds for rearing imported fish, and on a European tour in 1852 Allport witnessed the passion for salmon angling in Scotland and Wales, and could see the 'social class' appeal for his circle while coarser tench and perch could suffice for those without angling experience or the ability to afford salmon licences.

Not lacking confidence, Allport wanted to be the ambassador populating the island's rivers in his broader pursuit of a reputation as 'par excellence, the great naturalist of Tasmania', as pioneering writer-artist Louisa Meredith later described. He cultivated relationships with an old world fascinated by Antipodean exotica, sourcing and exporting specimens, not least thylacine (Tasmanian tiger) remains, and various fish species. He also traded Aboriginal skulls and bones, particularly to the Museum of the Royal College of Surgeons in London. In return, he became a fellow of Britain's Zoological and Linnean societies, member of several Belgium and French scientific societies, and 'Birdman' agent for distributing the famous bird studies of John and Elizabeth Gould, who spent much of a two-year colonial stint in Hobart. Gould was one of many who felt that 'had the rivers and seas of Tasmania been fitted for the *Salmonidae*, we should have found them there already', but Allport believed imported new species could acclimatise and overcome local deficiencies.

While Youl might have fretted he could be beaten on salmon, especially by Allport, but he received an unexpected fillip from a Victorian visitor. Edward Wilson, the son of a Nottinghamshire farmer, hadn't been able to emulate Youl's golden fleeces with his own sheep efforts, but with business partner Lauchlan Mackinnon made his fortune with the *Argus* newspaper in Melbourne. The paper boomed in Victoria's goldrush, allowing Wilson to pursue his passion for 'home' birdsong. As president of the Royal Society of Victoria's songbird committee, he transplanted lark, linnet, thrush, blackbird, goldfinch and nightingale.

He was in London for treatment for failing eyesight. But he also had a big vision as an apostle for acclimatisation, the new movement whose followers believed it was the duty of man to give the best of Creation's fish, plant or animal species to any country which could sustain them for profit or pleasure.

Edward Wilson saw Australia as a key player in a global exchange of species, with salmon 'most wanted'. Courtesy Victorian Collections.net.au

Wilson wanted to ensure Britain and Australia were not bested by rival France, where Napoleon III was offering rewards for those able to introduce, domesticate or acclimate species in his empire. He also wanted to know whether 'the production of fish by artificial means is more profitable than the cultivation of the land'.

Wilson aggressively promoted an 'if it lives, we want it' philosophy. In his mind, the British Empire would remain 'half-furnished' and unfulfilled unless it featured all possible species wherever they could be sustained. And its loyal Australian colonies, which had already succeeded with transplanting sheep and other animals, birds and flowers, were ripe for 'everything beautiful, profitable or interesting': 'Why should the tables of our colonists not be supplied with an occasional hare or pheasant as well as yours? Why should our alderman be baulked of his salmon cutlet, or slice of venison, more than your alderman? We do not know how many such things are capable of much more extended distribution than at present and we shall never know unless we try.'

To him, acclimatisation was 'one of the greatest causes upon which the human attention can ever be bestowed ... it promises to produce some of the greatest results the world has ever seen.'

Over dinner, Wilson and Youl shared their frustrations with the lack of home and colonial support for their passions, and how each had been ridiculed for their pursuits. Wilson could see the potential usefulness of Youl and his Association of Australian Colonies, while Youl was drawn to Wilson's energy and newspaper influence.

Neither was a true angler – 'not touched a fishing rod for twenty years', said Wilson – but they bonded on the salmon quest. They saw salmon, and possibly trout, as 'a great boon' to be bestowed 'into every suitable river in the colonies' of Tasmania, Victoria, New South Wales and New Zealand.

Salmon offered angling pleasure, food, and potential industry, but Wilson also argued 'a touch of poetic beauty'. So many institutions, traditions, characteristics and literature 'of our race' had been replicated in Australia, he said, that without a full transplanting of fish and birds 'how will our children comprehend ... the works of Chaucer, Ford,

Milton, Shakespeare, Spenser, Coleridge, Keats and 50 other poets?' Angling would also 'lead the Australian youth to seek their recreation upon the riverbank and mountainside rather than in the café and casino.'

Salmon also offered each the chance of some poetic revenge. Wilson had never forgotten 'a regular cannonade of rotten eggs' in Tasmania over some criticism in his *Argus* newspaper. He hoped 'by helping to get out this valuable fish to suggestively rebuke my assailants by the sort of scriptural nature of my revenge'. Youl was also tiring of attending meetings and hearing people say, 'oh, here is the fellow again who is mad about salmon eggs. He is only fit for the lunatic asylum', telling him he was 'wasting his time and money on a visionary scheme which could never be carried out'.

He could also turn the tables on those Hobart elites who besmirched his late father and family name, denied his mother a pension, and failed to suitably support his quest – and silence those who suggested an absentee tax on rich colonists who fled to England on the plea of procuring education for their children. 'Leave that alone,' he tersely told the *Examiner*. A successful salmon industry, perhaps contributing £50,000 a year (£5.7 million today, A$11.7 million) would hopefully 'relieve us poor absentees from much of the odium which has been cast upon us, and prove that we are not unmindful of that country to which most of us owe almost everything we possess.'

Wilson urged Youl and his association to get more serious about acclimatisation, with salmon the 'most wanted'. The impetus was timely for Youl, as he had competition. Allport tried to ship 120 small salmon, perch, tench and carp fry. Only five perch and one carp survived, but it was enough to be lauded by the *Courier* as 'the first lot of live English fish to the colony', overlooking the fact one of Youl's

Retired Navy Captain William Langdon, successfully imported the first tench in 1859, and was in pursuit of salmon. Courtesy Tasmanian Archives.

former Midlands Association colleagues, retired Navy Captain William Langdon, had successfully imported tench in 1859.

The salmon bid failed, but others were in pursuit. Former Youl neighbour, parliamentarian John Wedge, ordered salmon ova from Galway, hoping to place it in the South Esk. Some were contained with charcoal in a tightly sealed glass bottle, others intermixed with damp moss in a bottle capped by perforated leather. Wedge was 'not sanguine of the success' but vowed he 'would not give up yet'. And Ernst Marwedel, the Hanover consul in Hobart, told the Royal Society that on a visit to France, Professor Victor Coste had offered to supply ova from his Huningue breeding facility.

There was also more speculation of salmon shipments from the west coast of North America, with one ship captain suggesting a £5 incentive for landing a healthy male and £10 for a female. It was also known that Richard Nettle, superintendent of fisheries in Lower Canada, was fertilising and hatching salmon and trout and showing an interest in Australia. Two Ohio physicians, Theodatus Garlick and Horace Ackley, were delighting American fair goers with evidence of artificial trout bred from ova collected at the Great Lakes, and New Zealanders saw the shorter Pacific route as the means to best Tasmania in having the first salmon.

Transplanting fish across Pacific or Atlantic oceans continued to be seen by some as Jules Verne-like, not just man going to the moon but fishing on it. But amidst a flood of 'how do we get the salmon?' inquiries, a salmon rush was on.

6

LATITUDES & ATTITUDES

To be the first to get salmon across the 'line', Youl had to resolve whether the ice presented a final or fatal solution.

Wealthy Italians and the French had long impressed guests with ice and snow chilling their food and drinks. The rest of Europe scoffed at this 'excessive, effeminate' luxury, but a new ice age had dawned. At a display in the Strand, crowds of amazed Britons, who had only known ice scraped from ponds or the roadside, could read a newspaper placed behind a two-foot block of perfectly clear ice shipped from Boston by the Wenham Lake Ice Company. Many touched the ice to confirm it was not glass.

Food journals promoted the benefits of having pure ice 'within reach of everyone', to be used to cool as much as sugar sweetened. With Queen Victoria an early fan of ensuring meat, milk and butter never went 'off', ice for home and business quickly became a 'must have'. Anyone with £8

Heiress Angela Burdett-Coutts suggested freezing as the solution. Portrait, Alamy Stock Photo.

Professor Thomas Huxley, known as Darwin's 'bulldog', advised freezing would kill the ova. Photograph by Lock & Whitfield, V0026596, © Wellcome Collection.

or so could have a small 'icehouse', a wooden cabinet lined with tin or zinc and packed with a block of ice.

Youl understood keeping the water cold was crucial, but the line between near-freezing and freezing was fine. And some famous names believed freezing was the answer. One was the richest heiress in England, Angela Burdett-Coutts, who inherited £1.8 million (£194 million today, A\$406 million) from her banking grandfather at the age of 23. The enterprising philanthropist was a close friend of Charles Dickens and shared his interest in the mystique of Australia. When Edward Wilson's acclimatisation call-to-arms in the *Times* caught her eye, she promised £500 and made a suggestion: wouldn't it be easiest to transport salmon ova 'reduced to a dormant condition by being frozen'?

She was not alone. Famous Arctic explorer John Franklin and surgeon-naturalist John Richardson radiated wonder with reports of frozen fish coming to life when thawed. They reported seeing North American carp frozen into 'a solid mass of ice (but) if thawed before the fire, even after being frozen for nearly two days, the fish would recover their animation'. A French journal *l'Institute* also said some caught fish were exposed to such a severe frost that in separating them 'their fins and tails broke', but 90 minutes later they were put in water and 'to the astonishment of the spectators' many continued swimming.

Wilson was taken with the idea. Surely, he felt, salmon ova was exposed to cold 'below the freezing point' in Scotland and Ireland, and even colder climates of Norway and Labrador. If, as he had been assured, 'live fish may be frozen stiff ... and yet be restored to an active condition', wasn't it reasonable to infer a semi-animate egg might remain vital? Or that fish could perhaps be rendered torpid by extreme cold and go into a 'hibernating' state akin to animals such as bats?

The freezing notion had been debunked two centuries before by John Hunter, Scottish founder of pathological anatomy, who wearied of tales 'so well attested that we are bound to believe it'. But his proof of the falsity was buried in obscure publications such as *Philosophical Transactions*. Claims of prestigious explorers were more widely read and seen as credible, and journals continued to publish articles 'On the Resuscitation of Frozen Fish'.

Some prominent colonists in Tasmania accused Youl of being blind to freezing as the obvious, easiest and cheapest plan. Yorkshire-born Dr Matthias Gaunt, who pioneered Tamar Valley vineyards, asserted that in Scotland's icy winters 'the spawn must be in a state of complete congelation' so if Youl would 'pack the ova in ice' there could be 'no difficulty in keeping it in this state during the voyage out'.

Wanting to fully understand the impact of frost and ice, Youl and Wilson desperately sought out numerous scientific and fishing men. David Mitchell, who as secretary of the London Zoological Society had driven the development of the 'Fish House' in Regent's Park, told them conclusive answers were not to be found: 'Neither he nor any of the thousands of books on natural history ... would give ... any authentic information on the subject.'

Wilson was frustrated: 'There does not seem to be a man in all England sufficiently scientific to tell us whether the process of freezing would be fatal to the life of the egg. It is astonishing to find how little is known of this kind of experiment, and how puzzled are even the best authorities by questions of the simplest nature.'

Wanting to 'try the experiment', he pressed Youl. 'It would not only be cheap, but as it would be little more trouble to procure and freeze a million eggs than a thousand, we might within the next six months have the salmon safely established in every suitable river throughout Australia and New Zealand.'

But Youl resisted. He and a visiting Tasmanian friend, William Archer, a leading Royal Society member and fellow of the Linnean Society, sought the opinion of one of the leading naturalists of the day, Professor Thomas Huxley, who had visited Tasmania. The biologist-anthropologist, known as Darwin's 'bulldog', experimented and was absolute: 'Every one was killed ... freezing ova would kill it.'

Youl also considered the work of Dr John Davy, Inspector-General of Army Hospitals and younger brother of famous chemist Sir Humphry Davy, citing him as one whose opinion upon anything connected with the *Salmonidae* 'is the best that could be obtained in Europe'. To help resolve Charles Darwin's questions about fish breeding and distribution, Davy had tested the vitality endurance of salmon ova transported in various

William Archer, Youl supporter. Photograph, courtesy Tasmanian Archives.

combinations of dry and wet wool, sand, cotton wadding and moss, from the Leven and Dee to his home at Ambleside. He then exposed all to varying air and water temperatures and moisture in fresh and salt water. He hatched only a few ova, mostly the more advanced kept in green moss, but told Darwin his instinct about ova being transportable was probably valid. It led Darwin to presciently forecast that, with more fish experiments, 'geographical distribution would become, in my opinion, a very different subject to what it is now'.

Davy's experiments echoed the French experience. Naturalist-ichthyologist Charles Millet, vice-president of the pisciculture section of the French Acclimatisation Society, wrapped fecundated eggs in layers of wet sand or cloth in a pine box of moss inside a portable icebox. It was 'curious' to him that fecundated eggs of salmon or trout did not perish even when the cloth and moss itself became frozen.

Other experiments were underway across Europe, Russia, and America as men sought ways to transport ova between streams and ponds. Most declared 'nothing is easier to transport'. British Association biologist John Hogg agreed the day would come when spawn could be sent 'to other streams, however distant ... probably in time to any of our colonies in foreign climes'.

But there was a 'but'. All the experiments in Britain, Europe and North America were on land, with the longest journey 500 miles (800 kilometres) and none taking more than a few days – nothing like Youl's challenge of 16,000 miles (25,700 kilometres), three months of ocean buffeting, and huge temperature variations.

Davy 'looked rather grave' when Wilson pressed him on the Antipodean quest. Young fish of any age were not easily kept alive in agitated water, he said, and his experiments for Darwin underscored that high temperatures explained the absence of the *Salmonidae* in tropical seas and warmer waters. Even if ova could somehow survive the tropics, and even if rivers in Tasmania were cold enough, he feared summer sea temperatures at 40–42° latitude south would be fatal as salmon were 'never found in the Mediterranean' at 42° north.

Youl and Wilson fretted this might kill the dream of salmon surviving in Australian waters. But then *Field* reported that salmon had been found in the Bidasoa River separating France from Spain. As its 43° latitude was similar to Hobart, 'perhaps there may be less in Dr Davy's suggestion than we think'. It was also suggested salmon were plentiful around San Francisco at 37° north.

It was about attitudes and latitudes. Some said reaching for the grand prize of salmon was madness: freshwater fish too vulnerable; the voyage too long, rough, and hot; Tasmanian waters too warm and too full of dangerous predators. Many, including Robert Ramsbottom, urged Youl to trial hardier or non-migratory fish such as carp, tench, perhaps trout. Only if this proved successful should he try more delicate salmon. Wilson felt it would also give sea captains the experience of dealing with live fish before taking on 'the great king'.

But Youl only had eyes for salmon. Wilson realised Youl was 'a gentleman with a will of his own and having adopted the cause of the salmon with an enthusiasm which one cannot but admire, appeals in any other direction are only too often to be met by a dignified but austere rebuff.'

Youl's will was to discover 'the most effectual and cheap means' of delivering the Holy Grail without an unholy cost. Without committed funding from Tasmania, and nothing from the Home Government, he counted on a subscription drive. Led by William Wentworth, a former grazier who helped forge the first path inland from Sydney across the Blue Mountains, and Scottish-born merchant William Westgarth, a key ally in the anti-transportation movement, the target was £800 (£84,000 today, A$173,000). But they struggled to £650, well short of what Youl felt was needed.

It forced Wilson to relent: 'It appears more prudent to confine ourselves to the ova, and to avoid the proverbial hazards of him who seeketh to sit upon more stools than one.' Compared with the 'expensive, risky and troublesome thing' of large amounts of water for older fish, ova 'required little more than to be kept wet'.

Wilson also cited his godly view of acclimatisation: 'If we set deliberately about the task of supplementing Nature, we must imitate her in one of the most remarkable and beautiful of her attributes – profusion.

If she desires to keep a particular fish in existence, she provides each female with a roe of 10,000 eggs. And so, with us. If we are seriously desirous of extending among our fellow creatures the good things furnished us by a benevolent Creator we must not only be thoughtful, painstaking, and enterprising, but persistent, undaunted and profuse.'

When it came to keeping ova cool and unhatched, every degree of temperature and every day mattered. The Scots had experienced hatching not occurring for as much as 130 days in rivers almost freezing, and Youl was also told that if hatching occurred after 80 days, young fry could survive a further six weeks without feeding, by which time a ship should reach Hobart. If they did hatch and their embryo sac expired before landing, Professor Louis Agassiz and John Shaw suggested hanging carrion on the top of a pool to supply maggots, while Francis Mackenzie suggested droppings of cattle to deliver worms and insect ova.

Before thinking about the need for any maggots and worms, Youl wanted some ova experiments with different water and air temperatures to compare with records from the failed *Columbus* voyage. Crystal Palace approved some fresh experiments by John Pepper, a popular British science lecturer and star performer at London's Royal Polytechnic Institution.

Many warned it was too late in the 1858 spawning season to obtain sufficient ova, but Ramsbottom persuaded Youl that salmon spawned late in the Dovey River in mid-Wales. He duly collected 12,000 ova and transported it to Sydenham, where the eggs were placed in shallow wooden boxes, 24 inches long, nine inches wide and three inches deep (approximately 60x22x7 centimetres), placed inside larger boxes in a tank of flowing filtered water.

With ova in the tank and some subscriptions in the bank, Youl was feeling a tad optimistic. He told a friend in Launceston he had 12,000 ova 'for transmission to Tasmania (at) the first favourable opportunity'. And he was 'so thoroughly convinced of the value this fish would give to the colony that I would cheerfully give a cheque for £500 to have 10,000 ova placed in the North or South Esk River.'

This excited the *Courier* to confidently report 'they will be despatched for the rivers of Tasmania ... thence they will be distributed to the various rivers of Australia and New Zealand.'

But Youl wasn't over the line yet. 'To get them out is a very difficult matter. The heat alone … I dread, having ascertained that they cannot live in water of a temperature of 70°F (21°C) for a single minute. How we are to keep the water cool enough in the tropics without hundreds of tons of ice is a puzzle to me at present.'

He called for additional subscriptions from Tasmania, 'not to aid us in our first attempt but in case of failure to go on until we do succeed. We do not mean to let one failure stop us.'

Wilson told his *Argus* the 'great difficulties … do not seem insuperable … as soon as the young fry are hatched, and have got through their most tender stage, we shall take steps to forward them … the first consignment to Tasmania.' This suggested continued uncertainty, or conflict, over whether ova or hatched fry were to be despatched, but in any event, there would be no shipment. The Crystal Palace experiment was another miserable failure. The tropical department warmth, insufficient water flow and inadequate filtering of poor-quality Lambeth water caused fine weed to coat the ova. Then workmen mistakenly turned off the palace water for repairs. A few ova put in a fresh tank hatched, but none survived.

In a report to the Zoological Society, Bartlett nevertheless cited 'important facts learned from this experiment', including that it proved young fish could be hatched in as few as 30 days. As for delayed hatching, he merely noted it was generally impossible to slow ordinary birthing in nature without destroying mother or offspring. 'No positive conclusion can be arrived at without further experiments', he said.

In a prevailing climate of scepticism and insufficient subscribers, Youl was frustrated, describing the results of Pepper's 50 trials as merely 'interesting'. What he wanted to be more confident about was the Scottish view that hatching could be halted for 120–140 days in water just above freezing.

Pepper calculated that 12–15 tons (10–13 tonnes) of ice, packed in sawdust, would suffice for the journey to Tasmania, but Youl believed considerably more was needed: '25 tons the least quantity I thought we should require', perhaps as much as 'hundreds of tons'. But if such a large amount of ice was required it would add to the expense and the difficulty of finding a ship owner willing to take the risk of melting ice destroying other cargo.

The key question remained unresolved: how to preserve ova in an unnatural environment through unnatural heat for 100 days or more?

Waiting patiently for the next spawning season, Youl considered who should midwife a shipment of salmon brood across the seas. Admitting the 'dread' of not finding someone 'practically acquainted with the artificial breeding of fishes', he turned to Robert Ramsbottom. The potential reward for landing salmon was eye-catching, but Ramsbottom was approaching fifty, and it would mean leaving his family and propagation livelihood to spend months at sea and perhaps an extended period on the other side of the world.

Ramsbottom approached the Ribble's head bailiff, James Birch, to see if he was prepared to be involved. But Thomas Garnett advised Birch: 'Let him (Ramsbottom) make the first attempt' if he wished. He expected 'no great hopes', with the destination 'too near the tropics (to offer) a reasonable chance of success'. He had never heard of salmon below 45° north in Europe or anywhere washed by the Mediterranean, so it must be such climates were 'not congenial'. And he considered Tasmania a fatal shore because 'voracious and destructive' sharks would 'snap up every salmon or salmon fry', along with seals, porpoises, albatrosses, man-of-war birds 'and fifty other nameless enemies'.

Ramsbottom suggested his 24-year-old son William, who had been at his side for years, including with salmon propagation in Ireland. Youl immediately saw the merit, and the Colonies' Association quickly told the Royal Society in Hobart it was 'determined to engage the son of Ramsbottom, so well-known for his successful management of various piscicultural establishments, to bring out salmon or young fry to Tasmania'.

But it wasn't quite so straightforward. William Ramsbottom was on the other side of the world in Melbourne. He had perhaps grown weary of the Lancashire doldrums and didn't fancy spending more long days and nights in freezing water to catch and milk slippery fish or run a tackle and outfitting shop.

After marrying Elizabeth Catlow, an attractive 23-year-old cotton loom worker in September 1857, William was in Liverpool three days later, alone, to board the clipper *Messenger* as an unassisted migrant. He was part of an exodus wanting to see what Australia might offer he and his bride, who would not depart for another year. He arrived in Melbourne's December heat amid goldrush fever, along with *Messenger's* cargo of boots, shoes, railway iron, buckets and barrow wheels, and 250 boxes of pickled fish.

William could not imagine he was destined to one day be associated with a much more significant fish consignment, but that was for a day to come. As much as Youl wanted otherwise, it would take at least three months to get a letter to someone in Melbourne, and even if they found Ramsbottom it would be another three months to receive any reply. That risked being too late for the next spawning season and a shipment, when Youl had already spent five years on his quest and others were in the salmon race.

And Youl had already told the Royal Society, somewhat ambitiously, that it should ensure it had 'a place ready for the young fry as soon as they arrive in Hobart Town or after overcoming the dangers of the voyage they may be lost in your harbour'. Youl was not entirely convinced he could halt ova hatching for the entire voyage, or long enough to allow hatched fry to survive on their own embryo sac. But he remained convinced 'the best and most likely way to ensure success was by the ova, and not with young or old fish'.

Others maintained he was misguided. Robert Ramsbottom had an 'adverse opinion', as did Sir William Jardine, the influential Scottish naturalist, who declared it 'impossible' to carry spawn across the equator. But Youl ignored what he later called 'the ridicule or reasonings of other pisciculturists' who lambasted 'this act of folly, as it was called'. None, he declared, 'could persuade me to give up my own well-considered plan in deference to theirs'.

He just had to find the right midwife for a voyage, the right 'nursery' apparatus, and the right vessel. And quickly.

7

A LEAP OF FAITH

The salmon reward of at least £500 (£52,900 today, A$110,000), five times what many workers earned in a year, was the talk of Britain. In Liverpool it caught the attention of a Scottish sailor, Alexander Black. he sought and received the Royal Society's reward details from Youl, the two met in London.

Black claimed experience in 'many experiments ... in various parts of the world' to discover the best means of shipping birds and the ova of fishes, including salmon, through different climates. Whether it was true, in the absence of either of the Ramsbottoms, Youl accepted Black as 'a most fit person'. He told the *Liverpool Mercury* he had secured 'a person accustomed to salmon ova preserving', and when the news was relayed to Edward Wilson his Melbourne *Argus* cited Black as 'an experienced, enthusiastic and painstaking man' chosen for his 'practical acquaintance' with the subject.

The terms were somewhat uncertain. Privately, Youl considered any Tasmanian reward would first reimburse him and his fellow London subscribers. He merely assured Black that, if successful, he would be 'most liberally rewarded, and the value of his public services recognised with no grudging hand'. The *Glasgow Herald* correspondent said, 'so anxious is Mr Youl for the success of the experiment that out of his own pocket he offers the superintendent (Black) a premium of £100 if he succeeds in landing 10,000 ova in vital condition, and a greater or less sum in proportion to the number saved.' But in a letter to the *Examiner* in Launceston, Youl said the £100 would be divided between Black, the captain and crew.

The next task was to find a suitable ship. With the usual journey to Australia of about 120 days too long for delicate ova, many urged Youl to secure the fastest possible ship, preferably sailing direct to Hobart. His eyes first went to Isambard Brunel's famed 322-foot (98-metre) steamer *Great Britain*, which regularly made the run from Liverpool to Melbourne in 60 days. But its freight charge was too expensive for Youl, and the owners wouldn't countenance it sailing via Hobart.

Youl couldn't find an owner willing to sail to Hobart with affordable space for 20 tons (18 tonnes) of ice and overcome fears of melting ice damaging other cargo. In desperation, Wilson brazenly suggested to the Colonial Office that given 'the magnificent uselessness of the Royal Navy in times of peace' a naval vessel could be utilised. Lord Newcastle was horrified but dutifully promised to relay the notion to Queen Victoria. As she and Prince Albert had 'evinced a great interest in what was being done', Youl thought 'there is some hope' that Her Majesty's Navy might come to the rescue. But the issue of transporting fish was quietly left to individual commanders to ponder in the fullness of time.

Meanwhile, Youl put ova collector Robert Ramsbottom on standby while he considered a nursery apparatus to keep ova cool and secure. He recognised it would be 'very difficult to manage running water in a rough sea', and the smallest movement or contamination could instantly destroy ova by the hundreds, even thousands. He had in mind a more complex apparatus than that used by Gottlieb Boccius on the failed *Columbus* voyage, or versions sketched by the Scots and others in *Field*.

To replicate Nature's constant stream of clean, cool, oxygenated water, he envisaged 2000 gallons (9000 litres) of water in casks, a 200-gallon (900-litre) tank on deck and an icehouse capable of holding enough Wenham Lake ice to keep water cool for three months. The plan, or hope, was that perfectly pure water would continually flow nearly 100 feet (30 metres) from the tank on deck through the icehouse twice to cool, and then below deck to a cabin 'nursery'.

Here water would pass across ova embedded in small stones and gravel in trays inside shallow wooden boxes, similar to what Ramsbottom used in Ireland and Scotland. About a foot wide and six inches deep, they had perforated holes, 'a little less than peas', to allow water to pass. To keep the ova secure from pitching and rolling, the boxes would be suspended from the ceiling in a framework of gimbals and a series of chains, swivels and pulleys and stays. This echoed a 'hanging like a cabin light' proposal of James Burnett in Hobart after the *Columbus* failure. Newspaper reporters described the plan as 'somewhat complicated ... a curiously arranged ... combination of ingenious contrivances.'

Unable to find a suitable ship from London to Hobart, Youl turned to vessels sailing to Melbourne. These offered him more 'accommodation ... hence the chance of success greater' and for less cost, even if the salmon cargo had to be on-shipped to Hobart, where locals anticipated utilising North West Bay River, about 14 miles (23 kilometres) away, as a nursery. When this pursuit failed, desperate not to lose another spawning season, he told Black to inspect vessels in Liverpool.

White Star Lines promoted its clippers sailing to Melbourne twice a month as 'the largest, fastest and handsomest in the trade'. In berth was its largest vessel, the American-built *S.Curling*, which had previously served as an emigrant ship between Britain and North America. The 250-foot (76-metre) vessel was advertised to leave on 20 February 1860. Youl agreed to terms negotiated by Black and headed to Liverpool to 'take the management'.

S.Curling, an American-built migrant ship, Youl's first attempt.

But with White Star's promise of 'unswerving punctuality', everything was rushed. Ramsbottom had just nine days to find and propagate up to 50,000 salmon ova. And Youl, not happy with the quality of Liverpool's water, insisted he also secure 1800 gallons (8000 litres) of 'pure water' from a Clitheroe spring and train it to Liverpool in wooden casks. Only three days were allotted for building the icehouse and ova apparatus, with joiners and plumbers, unfamiliar with the task, working day and night.

Black was not a happy sailor. He recognised Youl's 'great exertions to procure ... the first importation of salmon ova' but was frustrated by what he diarised as 'interference ... procrastination ... mismanagement'. Youl abruptly adopted and abandoned ideas, he lamented, and demonstrated a 'niggardliness ... not a man to give more for an article' than he had to. While Youl thought 25 tons of ice was needed, he ordered only 15, wooden water barrels risked acidity but were not charred, and the icehouse was lined with lead 'as a cheap substitute'.

Black worried 'the whole thing was in imminent danger of falling through', at risk of being 'defeated by caprice, ignorance, delay and negligence'.

Youl had also been forced to initiate a second subscription drive to meet the anticipated spending of 'at least 600-700' pounds (£67,300 today, A$130,000), less than half what he first thought necessary to ever be successful. He was also inexperienced in managing such a project, and later conceded 'everything had to be hurried'.

With the clock ticking, Ramsbottom travelled 150 miles (240 kilometres) to the River Dovey in Wales to secure salmon ova. Youl became 'extremely anxious' when he had not heard from him for several days. Surely S.Curling would not depart without his precious cargo? He was much relieved when Ramsbottom finally arrived in Liverpool. He had only 30,000 eggs but Youl was pleased to learn of 'not half a dozen bad eggs in the lot'. Liverpool authorities permitted the ova being temporarily stored in large flowerpots under running water in the city's zoological gardens.

Youl telegrammed William Henty, the Colonial Secretary in Tasmania: 'I am off to Liverpool to ship 30,000 salmon ova, just obtained for me in Wales; they go to Melbourne, please make arrangements to receive them'. Henty hoped the next mail 'will bring fuller advices', but in case the ova arrived before the next mail he asked Ferdinand von Mueller, the government botanist in Melbourne, to spare one of his 'servants' to take charge of the ova and take it to Hobart.

When *S.Curling* Captain Frederick Gilchrist made it clear they would depart on 25 February, ready or not, all was not ready. There were still issues with the icehouse, water piping and charcoal purification, no time to fully test the swinging nursery bed, and concerns about the amount of ice. 'Our calculation,' Wilson wrote, 'is that with the most liberal expenditure of ice that we can afford, we shall have great difficulty in keeping down the temperature of the water through the tropics to about 45ºF (7ºC) (but) I feel sure that Mr Youl will have expended the last halfpenny available in adding to the supply.'

When a tug pulled *S.Curling* into the harbour, two joiners and a plumber were still working feverishly until, with a storm brewing, Captain Gilchrist told Youl, Ramsbottom and the workmen they needed to depart with the pilot.

Dr Ferdinand von Mueller, director of the Royal Botanic Gardens, promised to 'not leave anything undone' in securing the safety of the salmon ova or fry. Photograph, P-215.3. Courtesy Royal Historical Society of Victoria.

Youl farewelled Black with firm instructions to 'keep a journal of everything connected with the experiment for future guidance', a request he came to regret. *The Journal of Mr Alexander Black, in charge of salmon ova, in the S.Courling* [*sic*] would comprise details on expenditure, ship choice, apparatus and ice. Black diarised that very day he had 'expressed my resentment' to Youl about the mission management.

Despite the tension, Youl publicly evinced hope. He admitted still having 'grave doubts' about transporting ova until 'I saw our apparatus in working order in the *S.Curling* ... everything worked beautifully, and to the satisfaction of the old fisherman (Ramsbottom) ... I shall not easily forget his exclamation that if he were forty years younger, he would go out with them himself.' He confidently told Wilson 'our great effort would reach its climax'.

British publications had long marvelled, or sniffed, at colonists' efforts to transplant so many reminders of home, but this, the Glasgow *Herald* described, was a 'still more singular adventure ... a leap of faith'.

Youl's faith was quickly tested. A southerly gale turned the Mersey River into 'a boiling cauldron' and the vessel narrowly escaped serious damage when it collided with a French vessel, laden with brandy and wine, on its way to Liverpool. The *Morning Chronicle* reported it was many years since 'such ... destruction as the fury of this continuous storm has caused'. Gales blew two dock labourers into the Grand Surrey Canal, a workman off a scaffold in Berkeley Square and a child under a horse and cart. Streets were strewn with fallen chimney stacks, sparking fires, hospitals busy with people badly injured by falling debris.

Youl's heart sank as he read newspaper headlines of 'The Gale', 'The Terrific Gale', 'Alarming and Destructive Gale' and then 'The Hurricane'. Numerous vessels were in trouble with many a 'total wreck', including a steamer on its way to Cork with the loss of all 45 aboard.

Out in the boiling seas, in the first two days of a 13-day run of 'violent motion', Black recorded the ova toll. 'Should this weather continue long all the ova will die.'

Youl feared the worst for the ova – 'I dread (the storm's) effects upon our apparatus' – and his stewardship. He wrote irritably to Tasmanian newspapers to complain about the government's £500 reward and terms 'altogether inadequate' for someone 'voluntarily engaging in this work'. He and Wilson couldn't understand what they saw as a tiresome 'game of shuttlecock of responsibility between public support and private enterprise', when salmon in Tasmanian rivers could produce, for little cost, valuable food, an 'innocent amusement to the people', an attraction to visitors, and a valuable export to other colonies.

Perhaps sensing the risk of failure, Youl astutely abandoned any personal ambition for the reward. His salmon committee, he said, had 'directed me to make a present of the shipment ... without any claim for the reward'. He maintained 'I was always for this course' as it would be unfair for London colonists to gift alpacas to Victoria but take a different approach with salmon for 'the poorer government of Tasmania'.

Two weeks after the *S.Curling* departure, he wrote to the Royal Society advising a full letter was on board but said it should be ready 'should any number of the ova or fry alive in Melbourne'. If so, he called on it to charter *S.Curling* to Hobart 'as it would be a thousand pities to lose them by moving them from one ship to another'. He had spoken to the ship's agents, and they were prepared to do so 'at a small cost'.

The Society was again taken aback. Members like Morton Allport saw themselves as custodians of the salmon project, yet Youl had seized the initiative, taken two weeks to advise the *S.Curling* had left, and assumed last-minute Society funding of any trans-shipment to Hobart.

Meanwhile, Black was fighting a losing battle. Rough seas caused the ova apparatus to swing dangerously, making it impossible to change putrid water. A broken skylight caused 'my cabin float' and put salt water amongst the ova. Black diarised that he 'dared not use' the Clitheroe spring water as the uncharred wood casks made the water acidic, the ship's tank water was 'useless' because of iron oxide, and melted water from the icehouse was impure from its lead lining. He resorted to breaking ice into small pieces and washing it as best he could before laying it to melt on the gravel holding the ova. Whenever his thermometer showed the temperature rising to 39°F (4°C), he siphoned water off and laid fresh ice.

By the time the storms passed, so had most of the ova.

Survivors now faced equatorial threat. To the west of Cape Verde in the central Atlantic, Black's thermometer showed the temperature rising each day through to the 80s (mid-20s Celsius). 'Miserable about the ice as it diminishes very rapidly ... the heat is intense, and the passage long', he wrote. Made longer by what he saw as 'gross mismanagement' of the captain and poor seamanship of the crew.

But he drew some satisfaction on 2 April: 'The salmon Rubicon is crossed ... we have crossed the line.' He had transported some salmon eggs across the equator into the Southern Hemisphere for the first time. With favourable 'roaring forty' trade winds looming there was still a chance a few ova might reach the finish line, but then *S.Curling* hit the equatorial 'doldrums'.

Of the original 30,000 ova, only 239 now remained. He desperately put some in bottles of ice within wet blankets. They lasted the longest but nearing 30° latitude 'the ice is gone', so too the ova. It was 58 days and 15 hours since the ova came on board, 67 days and 15 hours since they were propagated by Ramsbottom.

It would be six months before any news of the *S.Curling* outcome reached London, but the fearful weather meant Youl felt he had to prepare the colony for the possibility of a second failure. On 18 April, just as Black was journaling the ova demise, and Tasmanians were receiving the exciting news *S.Curling*'s salmon was on its way, he penned an open letter to 'my fellow colonists':

'Notwithstanding all the precautions taken to insure [*sic*] success, the attempt to introduce salmon into Tasmania by the ship *S.Curling* may fail,' he wrote. But it was vital for the colony to hold its resolve and not forget the 'advantages likely to accrue from the introduction of this fish into their rivers'. Thomas Ashworth, the Galway fishery owner, told him salmon fisheries' annual value was £350,000 in Ireland, £600,000 in Scotland, £1,000,000 in Norway (approximately A$77 million, A$132 million, and A$219 million today). These were 'startling' results from introducing fish into rivers where they were previously unknown.

He hoped colonists would 'throw off the apathy' about the prospect, which he could only put down to 'either ... ignorance of the magnitude of the value to be derived from the introduction of salmon, or that they are so absorbed in attempts to discover a workable goldfield, that it has escaped their attention.' What he called the 'insatiate thirst' for gold irked Youl, especially with thousands of pounds being canvassed as a reward for discovery in Tasmania. To him it was a 'blighted and false hope' to think gold could be a permanent £300,000-a-year industry like salmon (£30 million today, A$62 million). The colony would be more sustainably enriched by new gold nuggets of salmon ova long after any goldfield was exhausted.

He was willing to persevere with salmon, but 'it is not to be expected that I can go on devoting my time and money to benefit those who will not lend a helping hand themselves. To the shame of my fellow colonists ... not a single shilling has been sent direct from Tasmania to help us. I have now by me nearly half a bushel basket of letters and papers on the subject, not more than half a dozen of which are from Tasmania, and only two of these offer any encouragement to anyone to proceed further.'

Youl concluded: 'Trusting these observations may be received in the spirit in which they are written, to increase the resources of our beautiful island. I remain your obedient servant.'

In a postscript he admitted that for a time 'I had very grave doubts' about being able to convey ova to Tasmania, but now 'I have none'. With some changes, such as 'more ice than we could afford ... the accomplishment was only a matter of expense'. No one could expect salmon to be delivered without 'much care, trouble and outlay'.

The pending arrival of 'the salmon ova for Tasmania' was anticipated throughout Australia. Anxiety rose when strong winds whipped up seas off the southern coast, and a ship from Russian Finland was wrecked off Kangaroo Island, with the captain, nine men and all cargo lost.

Finally, the telegraph station at the fishing village of Queenscliff, messaged Melbourne on Friday 8 June: 'The *S.Curling*, with the salmon ova on board, has just anchored in the south channel.' In a few hours, newspapers said, everyone would know whether 'the hopes of stocking the Australasian waters with the monarch of the rivers are for the present ill or well-founded'.

Waiting on the crowded Hobson's Bay pier was botanist Dr von Mueller, director of the Royal Botanic Gardens, who had promised to 'not leave anything undone' in securing the safety of the salmon ova or fry, 'should they arrive in vitality'.

He had organised 16 large casks containing 2000 gallons (7500 litres) of water for a temporary brick 'reservoir' beside the Yarra, with ice from the North Melbourne ice works of James Harrison, who would become known as the 'father of refrigeration', and from Adelaide. He ordered another cask of mountain spring water from the Dandenong Ranges in case the ova or fry were too precarious to make it to Hobart and needed to be detained in Melbourne. The botanist thought some of his plans possibly superfluous, but he wanted to 'not leave anything undone' for managing either eggs or small fry before the Tasmanian Government took over with its agreed chartering of *S.Curling* to North West Bay River, outside Hobart.

The German-born von Mueller was joined by two natural scientists: Irishman Frederick McCoy, one of the first professors appointed to the new University of Melbourne, who led acclimatisation efforts and particularly sought English birds to enliven 'the savage silence' of the Antipodes; and Paul MacGillivray, a prominent Scottish doctor and member of the Linnean Society.

With them was William Ramsbottom, described by newspapers as 'the son of the celebrated British fisherman, and by him fully initiated into the art of salmon propagation'. He had followed newspaper reports of the salmon quest, his father's involvement, and Youl's desire to engage 'the son of Ramsbottom' during his two years living with Elizabeth in a rented two-roomed brick and shingle cottage in Prahran. He had become a father to baby Jane, named after his mother, and was enjoying life teaching at a private school in the gold-fuelled rise of 'marvellous Melbourne'.

Now engaged by von Mueller for twelve shillings a day to help with preparations, arrival and transfer, Ramsbottom's eyes were drawn to other extraordinary and

William Ramsbottom, caricature, John Manly, *Salmoniana*, 1866

historic exotica, or what some dismissed as 'more absurdities'. Twenty-four camels from India were being unloaded for Robert O'Hara Burke's ill-fated expeditionary bid to win a prize for being first to cross the Australian continent from south to north, and fulfil von Mueller's hopes they would help 'conquer the desert'. No one could foresee that these camels were the origins of a family tree which would climb to more than a million feral camels. And the first five English hares were being unloaded for William Lyall, a founding Acclimatisation Society member, six months after fellow Society member and former Tasmanian Thomas Austin released 13 imported rabbits on his property at Winchelsea, introducing the fastest growing introduced mammal pest in the world.

The prospect of salmon history was revealed after 104 days of trepidation. On 9 June 1860, the anxious scientists, Ramsbottom and a waiting crowd learned the result was not historic, but horrific. The ova were dead on arrival, just like the *Columbus* failure. Newspapers called it a 'disaster' and 'public calamity' that this 'expensive, promising and patriotic experiment' had seen the precious ova 'doomed to find a dishonoured grave in the Atlantic'.

The Hobart *Mercury* did not directly name Youl, but its target for the 'full crop of ... disasters' was clear. The 'management' of the mission had failed to deliver 'a new industry as well as gratify the palate with a new sensation'.

Dr von Mueller was not so downhearted: 'It is ... gratifying to learn by this first experiment that salmon ova can be shipped to these shores, by such modifications of the apparatus as this first trial suggested, and by the copious and sole employment of ice for irrigation.' He hoped Tasmanian and other colonists in the Southern Hemisphere would learn from the experience and ultimately enrich their rivers as a lasting benefit to future generations.

But after two failures, sceptics declared the failure vindicated their view that 'the mad attempt' would never succeed. It was, one said, 'contrary to nature, and what is contrary to nature the wisest and wealthiest cannot accomplish'.

8

SILVER LININGS

Salmon dreamers looked to any silver lining. Some ova had survived 60 days and lived to cross the equatorial line to a few degrees south of the Tropic of Capricorn. Keeping eggs in a living state for about the time a more suitable ship might make the journey was seen as progress.

Edward Wilson, travelling in Europe, was quick to argue that 'mortifying as the intelligence is, it is not all bad'. Getting ova alive to 26° south 'is in itself a triumph and it shows the practicability of the scheme itself. Another ten degrees would have saved them.' With the precious eggs perhaps within a few days of latitudes where there would have been less need for ice, he felt, the quest ought not be lost 'for want of a ton or two of ice ... a little more ice, a little more experience and the thing is done'.

His Melbourne *Argus* recalled the legend of Scottish King, Robert Bruce, who famously took 'try again' inspiration from a spider which failed repeatedly before it succeeded in spinning a web, stirring him to deliver victory over the English. Wilson saw the history of British enterprise as 'replete with instances of lamentable failures preceding brilliant successes'. Salmon success was 'simply a question of expense'.

Despite the failure, the *Mercury* felt spending a few thousand pounds was now justified. 'We are no longer acting in the dark in the matter.' Even with rushed arrangements, insufficient ice, apparatus defects, fierce storms and equatorial delays, it was proved possible to bring impregnated ova through and out of the tropics.

Success seemed certain 'with more perfect arrangements ... properly observed, so a few thousand pounds would not be thrown away'. It would be more economical to be liberally 'doing the thing at once than a dozen conducted with a false and [miserly] economy'.

The Launceston *Examiner* agreed the enterprise ought not be abandoned. Thousands of pounds seemed a lot, but '[stinginess] ... is certain to end in disappointment and regret ... if the experiment is worth making it is worth making well.' It supported Youl's frustration about his efforts passing 'unsupported and almost unheeded'. The local indifference was akin to ignorance, 'just as a savage fails to appreciate the purest gem'.

In Portland, where Youl's friends, the Hentys, had pioneered Victoria's first permanent European settlement, the *Portland Guardian* editor, who also had spent time in Van Diemen's Land, urged 'disciples of good old Izaac Walton ... prepare your tackle!'

While this was perhaps a knowing nod to Youl's character, and there was broad support for another attempt, everyone had an opinion about 'how' and 'who': whether it should again be salmon ova, or young fry in salt water, as Dr von Mueller recommended to London; or two-year-old parr; or, as some continued to argue, by freezing spawn and young fish. And whether the Colonies Association, driven by Youl, ought to be given another chance, or whether the salmon reward should also be publicised in California and British Columbia for a less 'tedious and trying' solution.

Pointedly, the *Mercury* said, 'friends at home' ought not take any alternate salmon ideas 'ungraciously'. All ideas 'must be taken in hand by men who feel that if they fail all must fail'. Men of zeal and energy, meaning Youl, were utilising 'the most approved science and best practical knowledge' but the path to success could come from another direction.

But, on balance, it favoured another experiment from British waters. The support of noblemen and gentlemen in the 'mother country' could not be lightly thrown aside, and the colony could not afford to dispense with the services of 'zealous colonists' devoting themselves to 'the patriotic enterprise ... with a noble ardour to the task of peopling our woods and rivers with the feathered and finny tribes that furnish so many of the table luxuries of the Old World.'

Youl was one such zealot, and he was also actively engaged on other island agendas, including the recruitment of female emigrants. He was seen as 'sensitive' about the lack of appreciation for his efforts, but the *Mercury* conceded that if the government did not fully support the salmon quest 'we shall deserve the reproval of the world for casting our opportunities away'.

With funds in London 'exhausted', Wilson's *Argus* hoped news of more than 'paltry' support from all colonies could accompany the reports of the *S.Curling* failure. This would ensure 'private but patriotic individuals ... expending their time and money on our behalf' would not give up for want of local support.

The failure was reviewed by a Tasmanian parliamentary committee headed by Youl's friend William Archer, joined by another Youl supporter, grazier James Maclanachan, and two doctors from the Royal Society, Robert Officer and Henry Butler. Under Archer's guiding hand criticism was directed at the 'London committee', and the report judged that with knowledge of the reasons for failure, Alexander Black's suggested improvements, and more substantial funds, 'we may anticipate a more successful result'.

Black was seen by many in Hobart as the best man to manage the next shipment after what was described as his 'spirit of determined perseverance against the odds'. But Black insisted he would not take 'dictation or control of any kind' from such an 'irresponsible society' as the Association for the Australian Colonies, nor work for Youl. He wanted to be free to spend an agreed budget with 'no impediment, no interference'.

He sought a £500 fee, additional 'rewards' contingent upon success, and the assistance of two men on the voyage at £75 each. He also advocated that 35 fishermen - 'married men of good character and industrious habits' - brought out to run a local pisciculture operation, and regular importation of Wenham Lake ice if locals could not access ice from near the Antarctic.

The committee accepted his estimation that 50,000 ova would lead in a few years to a 'marvellous' 25 million fish, such that in six or eight years 'every river in Tasmania might be fished for salmon'. And agreed that the Derwent and tributaries were 'precisely such as the salmon would select', little different from Scotland's Tay and Tweed, and Ireland's Shannon and Laune.

But members, especially Archer, did not warm to Black's 'ungentlemanly' criticism of Youl, his terms, or his proposals. It resolved that supervision of another experiment be left with the Association in London and called for it to take advantage of the approaching spawning season and 'transmit (ova) by a fast-sailing clipper ship direct to Tasmania'.

The committee estimated it might cost £3000 (£311,000 today, A$648,000) for another voyage and establishment of breeding ponds. The government agreed, and Victoria contributed £500 on condition that the matter should be left to the association in London, because it could see placing management in Black's hands would be 'antagonistic'.

This was a victory for Youl, more so as the review committee also supported his view that even another failure 'ought not to discourage effort, since the difficulties which surround the execution of the project will enhance the satisfaction arising from success'. But there would be no progress in the upcoming spawning season. By the time news of the funding reached London, it was too late.

Rejected and dejected, Black tried to convince mainland colonies to engage him in their salmon ambitions, perhaps securing fish from France to place in the Snowy Mountains. Continuing to dispense his character references, he talked of 'underhand influences' striving to ensure the experiment remained in the hands of one 'jealous of the honour if not the reward of success falling to other hands'. He said Youl 'appropriates to himself the discovery of all … knowledge and congratulates himself' on inventions he had nothing to do with while solely blaming others for any failure'. And refused to pay tradesmen's bills 'unless they submit to the extortion … it is Youl all over'.

Back in London, a remarkable dinner helped energise the salmon movement. Three influential zoologists-naturalists gathered at the Albion Tavern to act as a 'tasting committee' to address what they saw as a failure to sufficiently broaden the British menu. Professor Richard Owen, superintendent of the British Museum's natural history department, David Mitchell, the first secretary of the Zoological Society of London, and Frank Buckland, a surgeon-natural historian, drew global attention for their exotic dinner of African eland, American partridge and European goose.

Buckland, described by friends as four-and-a-half feet (1.4 metres) in height and rather more in breadth, was already famous for being willing to dine on anything from anywhere since he was a child: hedgehog, field mouse, earwig ('horribly bitter'), ostrich, whale, porpoise, boa constrictor ('like veal') and kangaroo. His raffish showmanship and classical education made him a popular lecturer. Charles Darwin thought him 'incited more by a craving for notoriety … than by a love of science' but respected his mind.

And his mind was not just on spicing up Britain's staid diet, but also the need for more affordable food. Britain's population had exploded from nine million in 1801 to 23 million, and he was drawn to 'artificial fish' as the answer. As the son of the Dean of Westminster he said he felt compelled to do his duty 'in that station to which it has pleased God to call me … strive to become a mastermind, and thus able to influence others of weaker minds, whose shortcomings I must forgive.' To him, farming fish was like 'gold nuggets (which) have long been under our noses in the water, and we have not stooped to pick them up'. Fish could become a quality food supply for much less cost and effort than raising sheep and cattle, or African eland.

Frank Buckland, in a bid to enliven the British diet, drew global attention for an exotic dinner of African eland, American partridge and European goose, before he became a leading salmon advocate. Alamy Stock Photo.

Buckland persuaded *Field* owner, John Crockford, that fish could be the catalyst for a more united and successful approach to acclimatisation. Hunter-writer Grantley Berkeley duly wrote a series on the French Acclimatisation Society which had grown to a membership of 2000, including Napoleon, the King of Siam, and Emperor of Brazil, as a forerunner to the formation of a Society for the Acclimatisation of Animals in the United Kingdom. On the urgings of Wilson and Buckland, the society's first meeting in 1860 extended the remit to include fish. A network of 50 societies quickly spread, including Tasmania, which was built on acclimatisation foundations before the term was ever adopted.

Wilson and supporters saw it as Britain's manifest 'duty' to 'civilise' all corners of its Empire with species exchange because it possessed dominions in every climate and had unrivalled mercantile resources.

Charles Dickens and his wealthy friend Angela Burdett-Coutts were among those drawn to the rising movement. Dickens' *All Year Round* journal bemoaned: 'The cry of "Salmon in Danger!"' is now resounding throughout the length and breadth of the land.' Burdett-Coutts was taken with Wilson's questioning of whether it was right for the home government, 'surrounded by fifty-five colonies of unknown capacities, to leave everything to private enterprise?' And if private enterprise failed, whether it was right for noble undertakings to remain 'untested by any effort whatever?'

Youl was achingly familiar with divisions over public/private funding, and, following criticism of his salmon management, he now faced scrutiny over his efforts to support the government's desire for 'first-class' female emigrants. He struggled to identify hundreds of women without 'personal defect', with verifiable personal testimony, and a willingness to risk a long voyage away from family and every familiarity. He upset the Board of Immigration in Hobart by not waiting for approval of his request to utilise the aid of the St Andrews Immigration Society in Scotland, ignoring its views on numbers and capacity, and delivering some women seen as not unfamiliar with reformatory workhouses, prostitution and concubinage.

It was impossible for Youl to be certain of success with either female or finny emigrants, and on both fronts he felt his honorary work was not receiving tangible support, empathy or recognition. About to send to Launceston on the *Antipodes* what the immigration board had already warned would be a 'highly objectionable' number of English, Irish and Scottish women, and with salmon sceptics and critics baying, he needed to shore up his position by visiting Tasmania.

He chose the 'overland' route, where the Peninsula and Orient company offered twice-monthly voyages of about 50 days: a steamer from Southampton via Marseilles to British-held Alexandria; rail to Suez where work had begun on a new canal to cut travelling time to the East; another steamer to Aden and Point de Galle in the British colony of Ceylon (now Sri Lanka); another steamer to Melbourne and a final leg across Bass Strait.

After a seven-year absence, Youl reunited with his mother and siblings, and the *Cornwall Chronicle* welcomed 'an old and universally respected colonist, James A. Youl, Esquire,' and applauded his salmon exertions which 'will not soon be forgotten'.

Youl joined the mayor and clergymen in welcoming the *Antipodes'* cargo of women with a divine service and 'useful advice' before they were paraded for potential hiring. The *Examiner* said they were 'apparently of a superior class', but the rival *Chronicle* predicted the depressed economy would soon force or lure many beyond 'the path of virtue'.

After a family Christmas, Youl headed to Hobart to press his faith in the 'goodness' of emigrant women and salmon. He was greeted by the *Mercury* questioning why money was being spent to 'inflict upon us the foul thinking that tainted women are good enough for Australian wives and mothers'. England had become accustomed from the first, it said, to use Tasmania as an outlet for 'social evil', but if well-known and earnest colonists like Youl and their faithful emigration agent found it impossible to prevent some 'bad girls' from coming out, then the system was 'radically unsound' and should be 'forthwith abandoned'.

Youl protested it was impossible to prevent some 'bad girls' coming out or avoid being deceived, and 'it is most unfair to condemn a whole flock because of one or two conspicuous black sheep'. And he felt it unfair to be castigated for another honourable, difficult quest.

He was relieved his friend, Colonial Secretary William Henty, felt the same when he received Youl's female and fish accounts. Henty indicated the impending end of female emigrant shipments and reassured him the Colonists' Association in London should continue managing the salmon mission, for which he confirmed funding for another shipment and development of salmon receiving and breeding ponds.

Selecting a ponds site brought Youl into conflict with Morton Allport, who was determined that the Royal Society, through his hand, would forever be credited with having been 'the first to recognise the vast importance of this (salmon) undertaking ... and to take practical measures to ensure its success.' Allport was also offended Youl had not informed the Society of the *S.Curling* plans until after the vessel had left

England. And when his brother, Curzon, sought information in London on breeding pond requirements, Youl had given an assurance he had forwarded 'full instructions' to the Society but it 'never received such information at all'.

Allport had assessed possible sites for salmon ponds with the help of the head gardener at Government House, William Sangwell. Allport described Sangwell, an angler perhaps familiar with trout in his younger days around the River Kennet in Berkshire, as 'the only person practically acquainted with the modern science of pisciculture, that could be found'. Allport was initially drawn to the Derwent-Ouse River in the central highlands, well upstream from Hobart, but Sangwell favoured the closest site, North West Bay River at the head of D'Entrecasteaux Channel, twelve miles from Hobart. The government agreed and would engage Sangwell to construct some ponds.

But when Youl visited the proposed site, accompanied by friend William Archer, he immediately penned a letter declaring it 'not at all suited'. It was, he said, too close to the sea to allow young salmon to develop before entering salt water, too shallow with poor tidal river movement, and probably deficient in food. 'I would strongly urge that immediate steps be taken to provide ... another spot, higher up in the Derwent.' He favoured the Plenty River, a tributary to the Derwent, 28 miles north-west of Hobart, the same area promoted by his nemesis Alexander Black. Youl left his letter with Archer and quickly set off for Launceston, precluding any further discussion.

The northern mayor and 30 esquires feted and farewelled Youl with 'thorough English hospitality', feasting on fine wine and food and entertainment by 'Herr Wackeldine's artillery band'. Youl savoured toasts for his 'colonial loyalty and self-denial' in laying foundations for the colony's 'greatly improved social condition' through emigration, mail, and salmon.

After 'thoroughly hearty cheering', Youl declared salmon success 'a mere question of time and expense'. With knowledge gained from the S.Curling attempt and more government support, he had 'no doubts of eventual success'. But he couldn't resist raising the mooted absentee tax on those who had 'fled' to London with their riches: while admitting 'a little selfishness', he expected his salmon work to be rewarded by 'total exemption'.

On the P and O's *Behar* sailing from Melbourne on 26 January 1861, Youl reflected on his Antipodean summer. The journey meant another salmon attempt had been put on hold for a year, but he had more funding, management certainty, and had secured support for his preferred Salmon Ponds site. His hands were still on the salmon tiller.

9

SLEEPING WITH SALMON

Back in London, Youl had a long 'to do' list: helping showcase Tasmania at a world fair in London, organising a final shipment of female emigrants, and lobbying for reparations for the legacy cost of convictism.

But salmon was top of mind. Edward Wilson said 'he eats, drinks and sleeps with visions of the salmon in all its stages floating perpetually about him'. Sneering mutters of 'oh here is the fellow again who is mad about salmon eggs' continued, but others were beginning to see some method in the 'madness'.

When he first arrived in London in 1854, the metropolis was amazed to see fish exhibited in aquarium tanks. Now, William Lloyd, a self-taught zoologist and first professional aquarist, had hatched salmon in his shop in Regent's Park. Frank Buckland declared 'seven little beauties, all alive and kicking' as the metropolis' first recorded salmon hatching.

An article on 'Fresh Fish' in Charles Dickens' *All Year Round* journal also said principles of propagation were so clear it was being taken up by many for amusement or profit. But the journal cautioned promoters of propagation and acclimatisation not to make their 'sciences ridiculous by extravagant promises and visionary expectations which will be contradicted by the practical result'. Anyone who thought hatching fish was the whole act was making the same error as a farmer who considered his wheat-crop safe in a barn as soon as he saw green blades above the ground.

Youl understood that truth well enough, and that what worked in a London shopfront was a vastly different kettle of fish on the open seas. During his absence in Tasmania, two Colonies Association colleagues had also become fearful of another failure 'more discouraging and disappointing than the first'. But 'no one else has the smallest knowledge of or experience on the subject', so it had been 'imperfectly investigated … so little understood'. Chairman Sir Stuart Donaldson, a former New South Wales Premier, and secretary Lauchlan Mackinnon, a partner in the *Argus*, went to France to meet those who had taken salmon and trout propagation to a new level.

Professor Victor Coste, physician to Napoleon's Empress Eugenie, studied embryology and believed in the 'conquest of living nature'. He saw fish as the new realm where 'human industry' would organise 'the machinery of exploitation, where the fruits of this inexhaustible domain, held, ripened and multiplied with care, will be harvested with as much profit and less labor than those of the earth.' In an echo of Henry IV's mythologised promise in the sixteenth century of 'a chicken in every peasant's pot every Sunday', Coste vowed to supply Frenchmen with 'a daily giant trout'.

The director of the Huningue piscifactoire, in South-East France not far from Basel, advised Donaldson and Mackinnon 'the only mode' to any possible success would be to transport ova or fry on a steamship reaching Tasmania with the least possible delay. A steamship would allow them to 'carry as much ice as you can please' and reduce the length of a sailing voyage by half. Coste also strongly recommended trout be shipped as well as salmon, as 'they would almost certainly succeed even if salmon did not'.

His colleague, Zepherin Gerbe, an embryogenesis expert, explained how ova was transported all over France in moss-lined earthenware jars or wooden boxes. He gifted the Australians a perforated box to hold 1000 to 2000 ova inside a 'nest' of moss moistened with cold water. In turn, he said, this box could be put inside another closet surrounded by 'lumps of ice' so melting ice could drip through the box holes to the moss. This echoed the technique of Joesph Remy and others across Europe who transported eggs in fir, moss or aquatic plants inside two wet cloths, or fine wet sand, sometimes accompanied by a portable icebox. Gerbe also stressed ova be packed only after the emergence of tiny dark spots, signifying the emergence of eyes.

Mackinnon pressed Coste on any Antipodean success with latitudes corresponding with those around the mouth of the Rhone, where the Australians had been assured there 'were no salmon in the Mediterranean'. Coste replied: 'Ah! But there are salmon in the Mediterranean. I admit the salmon is not indigenous ... but I can affirm that there is nothing in the character of the waters, or its geographical position on the globe, to prevent salmon from thriving in it ... I myself have seen a full-grown salmon – seventeen pounds – in the river after it had evidently but just returned from the Mediterranean.'

But Dr John Davy, whom Youl had previously consulted, maintained such sightings were 'no proof that these fish habitually were to be found in that sea ... it is easy to conceive how a single stray fish might enter it by the straits of Gibraltar [sic] and wander about until it was caught.'

Youl wrestled with the unresolved. Whether ova needed a system of continuous running water or could survive in a moss-lined box with ice. Whether 'frozen' gravel, cloth or moss would kill the ova. And whether ova had to be sent as soon as possible after fecundation or only when the pigmentation of eyes began to appear.

He began to think the task might be beyond him: 'I despaired ... well-nigh given up'. But he was boosted by Irishmen, particularly Galway salmon pioneer Thomas Ashworth. 'He ... very earnestly urged me to ... persevere till I had placed this fish in (Tasmanian) rivers ... I took new courage.' Ashworth had conducted minor experiments on ova transportation, as did William Ffennell, an influential Irish fishing inspector-reformer, with fellow fishery commissioner Thomas Brady. Ffennell, who became Royal Commissioner of Salmon Fisheries of England and Wales, told Youl 'you must go to Dublin if you wish to learn anything about salmon'. Admitting he had to learn more, Youl embraced 'getting my first lessons on pisciculture ... to learn all I could on the subject' by visiting Ireland, then Scotland and France.

On the first of three visits to Ireland, he was reassured to see the familiarity of the countryside: 'I would not know it from Tasmania.' Brady shared what he had learned from his experiments and from Professor Coste and pisciculturist Charles-Auguste Millet. He educated Youl on every stage of the salmon's life and showed where hundreds of thousands of salmon had been propagated by the Ramsbottoms at Oughterard and Screebe.

On his trip back to London, Youl conducted a small experiment to see for himself how far fry might travel without ice. He placed six two-month-old fry in tin cans with air holes, and utilised additional cans of fresh water for replenishment. 'Notwithstanding all my care and attention, I only conveyed four alive to London, and one of them died the next day, a very strong proof of the difficulty of moving salmon fry of any age any great distance.'

He left Dublin better informed – 'I consolidated all I had read and previously seen on the subject' – and more appreciative of the need for better salmon preservation laws. He joined Ffennell, Frank Buckland and *Field*'s angling editor Francis Francis and others to form an Association for the Preservation of the Fisheries of Great Britain. This coincided with novelist William Thackeray lamenting in *Cornhill Magazine* that despite increased understanding of salmon's natural history 'the cry all over the country may soon be – WE HAVE NO SALMON'.

France's fish factory at Huningue was Youl's next classroom. Spurred on by Napoleon III's imperial ambitions, the hatchery was shipping tens of millions of eggs across Europe, bedded in moist moss or sand inside wood containers. Coste reaffirmed the 'safest plan' was to send ova in damp moss in a simple apparatus 'by a fast ship', the same advice he gave to Youl's colleagues, and earlier to Tasmanian Royal Society member Ernst Marwedel in 1856. Ashworth had also assured Youl that spawn of salmon, trout and other fish, had been 'successfully conveyed to great distances' in boxes filled with wet moss. From Sweden, Horace Wheelwright, an amateur naturalist who called himself 'Old Bushman' after spending many years in Australia, outlined to *Field* how fishermen transported ova from Norway to Sweden within moss kept constantly cool but not freezing and without running water. 'What nourishes them is the damp, chilly air they get from the moss.' Swedes had transported ova in juniper fir as far back as the 1700s, he said, and would not fear of taking ova to Australia as long as the voyage was not much more than two months, and the ova boxes were held in an ice container. Salmon ova, properly damped and managed at the right temperature, 'would travel around the world if the voyage were not too long'.

Youl accepted that salmon and trout ova packed in moss and other materials had been sent around France, Britain and Scandinavia, and in America around the Great Lakes, but the distances were short and on land. An extremely long voyage across the world and through extreme climate changes presented a more difficult challenge.

He pressed Gerbe on whether the moss method inside an icehouse would overcome a three-month journey to Australia. 'He most emphatically gave me his opinion that they would not live,' he reported, '... declared it impossible to send ova in moss on such a long journey.'

It left Youl convinced he still needed to create an artificial nursery replicating nature as closely as possible: to keep ova in a temperature not dropping below 35°F (1.6°C) or rising much over 50°F (10°C), in a constant stream of fresh, cold water, not too slow and not too fast. And somehow overcome the 'impossibility' of keeping the delicate ova constantly secure, as any sudden movement would be fatal.

Youl engaged Thomas Johnson, a young and intelligent ship carpenter, to develop models of apparatus with variations of wood, tin, zinc, lead, glass and different suspension joints, gimbals, universal joints, weights, chains, pulleys and elastic cords. Youl approved two models using similar principles, both to be suspended in a bid to offset the rocking and rolling forces which destroyed the *S.Curling* consignment. One was a pyramid of three tiers of trays, designed to swing 'to and fro' from a beam by a universal joint; the second a larger 'gimbal apparatus', intended to remain steady in the manner of a marine compass by utilising knowledge first used by ancient Greeks.

To imitate a natural spawning stream, the suspended apparatus would be fed with water running from a 400-gallon (1500-litre) tank, filled with charcoal and lined with pure tin, through a well of 25 tons (22.6 tonnes) of ice to cool before entering a smaller iron tank of 200 gallons (757 litres). Water would then be piped into a set of trays holding tens of thousands of eggs in a shallow layer of gravel. Water would pass progressively from tray to tray before being pumped back to the top to resume the process.

Newspapers lauded the approach as 'elaborate and ingenious', and with the winter spawning season looming, Youl was feeling buoyant. He planned to include the models in the Tasmanian section of London's forthcoming International Exhibition, and was pleased Francis Francis, Professor John Pepper, and 'many others ... all approve'. *Field* also felt some optimism, but with a big if: 'If enough ice and water can be conveyed, sufficiently far south, and the passage be only reasonably favorable and quick, we have very little doubt that the forthcoming trial will result in a success.'

It reported the Australians 'do not intend to place all their eggs in one basket', stating that Youl hoped to place ova and some small fry in up to three aquarium apparatus 'so that my chances of success may be multiplied', and 'if possible' to send ova in different stages of fecundation.

Youl took the two favoured apparatus on a voyage to Perth, the epicentre of Scotland's salmon exports. It was a connection with the home of family ancestors, and a reminder of Perth, Tasmania, where his mother now lived, but he mostly focused on seeing which apparatus best mitigated water from sloshing about. The marine compass version supported by gimbals 'we found to answer ... well'.

He met Tay fishery head Robert Buist and inspected the Stormontfield hatchery ponds. He sent to Tasmania's Royal Society a

Times outline of the ponds, which were 'simple in design, inexpensive in their construction and ... conducted at very little cost'. Two days later he sent a hurried note advising that Robert Ramsbottom had transferred 200,000 salmon ova to a district which had never had salmon and was now ready to supply Youl. This 'proves conclusively', he said, the need to quickly establish his recommended ponds site on the Plenty 'for ova or fry, should I succeed in sending any out alive'.

The 1861 annual meeting of the Colonies Association reported Youl as 'busily engaged' and with new information 'it may be confidently anticipated that the next experiment will be made under circumstances highly conducive to favorable results.'

The news reached Tasmania just as the government appointed a Salmon Commission to 'make arrangements for the reception of the salmon ova from England' and, the *Mercury* reported, to conduct 'the experiment in this colony for the naturalisation of the salmon'.

This followed a Royal Society report on the 'management' of any salmon spawn or fry to Tasmania. One of the authors and new Salmon Commissioner was Morton Allport, along with Royal Society secretary and Youl ally William Archer; House of Assembly Speaker Robert Officer; public works director William Falconer; and two men who had also pursued salmon, John Wedge, a former neighbour of Youl, and Captain William Langdon.

Allport and some of his Royal Society colleagues saw their role being broader than 'reception' and local management, but the government had already entrusted the next transplant bid to the London colonists. The Commissioners obsequiously declared they 'cannot but regard (the decision) as judicious ... but (the Association) immediately delegated to Mr Youl the sole superintendence of the necessary preparations for the renewed experiment about to be tried.'

The 'but' was not lost on Youl, who icily agreed to proceed. 'Feeling now confident of success, notwithstanding their abuse, and as I considered ... ingratitude,' he accepted the commissioners' 'trust'.

His next task was to engage a new salmon midwife. Following Robert Ramsbottom's confidence that his son William would 'readily return' to take charge of a shipment, Youl sent an urgent message to Edward Wilson in Melbourne, to find and engage 'the son of Ramsbottom'. Now 28, William had been in Melbourne for four years, witnessing gold seed one of the richest and fastest growing cities in the world. Its 125,000 people savoured a new University of Melbourne, National Museum of Victoria, State Library, Melbourne hospital, fine shops, pubs, gentlemen's clubs and theatres. Men talked passionately of what they called football and the running of the first Melbourne Cup.

But the promised £300 salary (£30,000 today, A$62,000) and potential reward of £300 to £500 was enough to persuade young Ramsbottom to board *Roxburgh Castle*, one of three so-called 'gold ships' taking millions of pounds worth to London. Wilson's news – 'I hunted him up and sent him home' – delighted Youl: 'I can hardly tell you how pleased I was to receive your letter ... telling me Ramsbottom had sailed.'

Notwithstanding some lingering local disquiet over Alexander Black not being re-engaged, the *Mercury* reprinted an *Argus* report that William had 'studied the whole subject from early youth under the guidance and instruction of his father, the celebrated fish propagator' and was 'eminently adapted to take charge of this precious consignment ... the great experiment ... if all goes well, the ova will leave England next February so as to arrive in Tasmania after the cool weather has set in, and we may now once more live in hope that by May next this invaluable fish may be safely placed in colonial streams.'

The 'if all goes well' caveat was soon underscored. The sighting of *Roxburgh Castle* off Cowes coincided with weeks of gales causing loss of life and havoc in the English Channel, and Ramsbottom could not disembark for another month.

Convinced that Youl had to secure a fast vessel able to withstand stormy seas, many again urged *Great Britain*, or Isambard Brunel's new record breaker, his 'great babe', *Great Eastern*, which could perhaps reach Australia in five weeks. Big steamers reduced the risk of an extended voyage and the lurching and rolling of a sail vessel, but Youl feared engine vibration would destroy ova vitality. And clipper owners remained nervous about melting ice and were not prepared to offer cargo space and insurance terms he considered 'reasonable' or adjust schedules to align with ova collection.

With the spawning season underway, Youl was under pressure and frustrated that he could not secure a ship because of what he dismissed as 'frivolous' excuses. Melbourne traders wouldn't hear of calling at Hobart first, and the little trade from London to Hobart involved vessels 'of the lowest class, slow sailers and not (with) sufficient space to make an efficient icehouse ... it is courting mischance to send the ova by one of them.' Then agents warned that with the Civil War underway, 'if we have a rupture with America, outward freight will rise considerably'. He sought a large vessel trading to New Zealand, such as the 1032-ton (936-tonne) clipper *Zealandia*, to call at Hobart en route, but the £750 price (£76,000 today, A$159,000) was too high for Youl.

Then a broker suggested *Beautiful Star*, in the final stage of construction at Newcastle-on-Tyne, destined to carry coal from Newcastle in New South Wales. Despite his 'courting mischance' concerns over small vessels, Youl was drawn to one of just 125 tons and 119-feet in length (113 tonnes and 36 metres) because he could have it exclusively for £500, paid in advance, with a gratuity of £50 'should 1000 ova or fish be landed alive in the colony'. He signed a Christmas agreement, on the condition it would 'go out under canvas only' to avoid any engine vibration from its two-cylinder screw engine. The vessel would leave Newcastle for London about 17 January for a final fit-out in London. Ramsbottom would mess with the captain, with an assistant, accommodated in steerage.

It was a big gamble. Tasmania's parliamentary funding was for 'a fast-sailing clipper ship direct to this colony', as the French, and some of Youl's association colleagues, advocated. The small *Beautiful Star* could not be expected to have a quick sail, and if a second mission failed it would be a falling star for Youl.

He knew it was risky. 'I very much wished the vessel was larger', he wrote in an explanatory letter to friend William Archer. 'I was most reluctantly compelled to accept (it as) the best available'. Compelled because of limited choices, tight funds, and his own impatience to be

Beautiful Star, Youl's second attempt.

first to land the miracle. 'I cannot tell you how anxious I have been and really began to fear the experiment would have to be deferred for another year.' He argued the vessel, 'small as she is (has) many advantages in her over an ordinary merchant ship.' It was sailing direct to Tasmania, which would avoid deferring the experiment for a second year and save hundreds of pounds. It had 'a great height' between decks to facilitate his nursery apparatus. And because he 'had the whole vessel to myself and plenty of time to put on board and ... try the workings of the apparatus before the ova is deposited', he could avoid a repeat of the rushed arrangements on the *S.Curling*.

But *Beautiful Star* arrived two weeks behind schedule for its final fit-out at London Docks, and with neither vessel nor apparatus completed it was getting late in the spawning season. And Robert Ramsbottom was busy pursuing fish for someone else. Italian military hero Giuseppe Garibaldi wanted salmon fishing at his Sardinian island, Caprera, so supporters engaged Ramsbottom to send 25,000 ova from Galway to Genoa. Ramsbottom proudly showed Youl a letter in which the Italian thanked him and hoped to get his advice as 'I am told you are so eminent a master'.

Ramsbottom assured an anxious Youl he still had time to deliver his ova from the Dovey. But he hadn't counted on fish felons. After securing two parent fish 'by a cord to their tails' two parent fish from the Dovey, he explained to the Mallwyd neighborhood 'the great object the fish were to be used for, and the enormous expense already incurred'. But during Sunday divine service, he reported, 'some scoundrel cut the cords and stole the fish'.

He captured two more spawn-laden salmon at Derwenlas which produced enough ova, for which Youl paid £49. With a £625 (£65,500 today, A\$137,100) apparatus and icehouse 'nearly complete', Youl told the *Australian and New Zealand Gazette* that anyone interested 'in this national experiment' was welcome to visit him at Wapping docks. 'I am here every day, I shall be very glad to explain the means adopted to ensure a constant stream of pure cold water being supplied to the ova during the long and perilous voyage.'

Visitors were intrigued by the intricacies put together in five weeks of 'incessant labour' by ship joiners, engineers, and plumbers. They could trace the basic plan of water running from one tank, sitting partly on deck, through an icehouse and into another tank before flowing over two contraptions holding ova inside gravel which William Ramsbottom had washed and boiled 'over and over' to remove any noxious elements.

Everything 'promised well', Youl wrote, but others were convinced that regardless of the apparatus merits such a small vessel as *Beautiful Star* could never complete the journey in reasonable time, perhaps not even within 160 days, and without fatal rock and roll. Dublin friend Brady offered Youl a suggestion. 'It strikes me,' the Fisheries Board secretary wrote, 'that you ought to try the ova in moss also.' He had recently received some ova in damp moss, and 'I never saw any ova in such good condition'. He thought it 'might keep so a very long time I think'.

'A small trial this way would do no harm. It can be easily watched to ascertain if they are coming to life. If they don't hatch before arrival, it will be a decidedly safe way of transporting them. I send you a sketch of what I would propose.' His drawing showed ova packed in layers of moss inside a box, similar to that proposed by the French, with iced water passing through the moss then drawn away. If iced water could retard the hatching, then a moss box 'I think will be the easiest way of preventing them being tossed about by the rolling of the ship as the moss will keep the ova steady'.

Youl was convinced it would not succeed because of the absence of running water, which was why he had made 'such elaborate and costly arrangements'. But he politely requested that Brady make a single wooden box to hold a few hundred ova. And as for a suggestion, by Wenham Lake Ice Company head Edward Moscrop, to bury the box in ice, Youl was barely interested. He believed running water was essential and that any freezing of ova would be fatal. The Ramsbottoms agreed, predicting the boxed ova would not last a fortnight.

With engines silent, *Beautiful Star* finally sailed in early March 1862 with about 50,000 ova, 25 tons (22.6 tonnes) of Wenham Lake ice, two nursery apparatus and one small wooden box. On board were Captain Charles Hodge, his wife Alice, a small crew, William Ramsbottom and an assistant, Mark New.

The whole world was watching, from the Southern Hemisphere out of desire for 'true' fish and perhaps a new industry, and in the Northern Hemisphere amid concerns that the lack of affordable food could trigger uprisings or war. President Abraham Lincoln had just ordered his Union forces to take 'aggressive action' against the Confederacy, favoured by British trading elites, and Europe was anxious about nationalist ambitions. An American Government report noted, 'happy would it be for nations ... instead of being exercised in feats of arms and destructive war-struggles, (they were) more frequently directed to ... the common good. The cost of one ship of war, or one regiment, how enormously does it exceed the expense of (the) beneficent establishment of Huningue!'

Youl's salmon dream was running in deeper waters.

10

A FALLING STAR

Farewelling William Ramsbottom on *Beautiful Star*'s maiden voyage, Youl had cause to reflect, ruefully, that tangible support for landing the finest English fish did not match another Antipodean obsession with English culture.

Having hosted social cricket matches at Symmons Plains and enjoyed the first inter-colonial match between Van Diemen's Land and the district of Port Phillip (now Victoria) in Launceston shortly before he left for London, his eye was drawn to an inaugural All-England Eleven tour of the colonies, the first visit by any sporting team. An earlier prospectus to help 'catch so many golden fish' failed, but two entrepreneurial caterers-promoters in Melbourne, in perhaps the first commercial sports sponsorship, garnered enough funds to offer each player £150, a free passage on the mighty *Great Britain* and all tour expenses.

Youl could only dream of similar support to land All-England salmon. The cricketers, who played against local teams of 22, were greeted with banquets and champagne receptions, and urged to regard their match venues as akin to moving from 'county to county'. In moving the Melbourne Cricket Club to make its ground available, barrister Butler Aspinall joshed that whatever the failings of introducing some species 'the game of cricket is readily propagated here'.

Nevertheless, Youl was optimistic, telling William Archer: 'Should the ova live through the tropics there is but little doubt of the success of the experiment ... when I left the ship at Blackwall everything seemed to work so well that success seemed certain.' But any prospect of a champagne-fuelled salmon reception was immediately at risk. Admiral Robert FitzRoy, the pioneering meteorologist, was issuing urgent telegrams warning of gales moving in from the Atlantic. The small *Beautiful Star* had barely entered the English Channel when it was forced to run back to Margate on the Kentish coast for three days. On the way it collided with another vessel.

Hearing the news, Youl immediately took a train to Deal. The vessel suffered little, but below deck there was concern: 'I found that one apparatus was a complete failure at sea, and would not act, and out of 12,000 ova placed in it not as many hundreds were alive.'

The swinging gimbal apparatus which 'worked so well' on his trial voyage to Scotland was not performing as well as the suspended trays. Youl admitted being 'very, very reluctant' to make alterations but relented as 'it did appear to be obstinate to adhere so pertinaciously to my own plans.' Ramsbottom made adjustments including Indian rubber stays for more stability.

'Perhaps', a hopeful Youl told a friend, 'these disasters in starting may have been the means of ensuring success'. But when the vessel neared the Isle of Wight, after not much more than a hundred miles (160 kilometres), storms again forced a return to the Downs before a third attempt on the night of 13 March. Then the loss of some propellor plugging and sail damage forced it to turn into the Isles of Scilly off the Cornish coast.

Those who later became familiar with *Beautiful Star* in its plying around New Zealand could not understand Youl's choice. 'She was a small, narrow boat,' author-angler Thomas Donne wrote, 'full of activity in heavy sea, and pitched and rolled in an extraordinarily unpleasant manner. She would sit on her stern and waggle herself, then try to stand on her head and waggle some more, vibrating until one's teeth played like castanets. Four or five months of this frolicking was enough to kill a man, let alone ova or fry.'

Some of its crew agreed. From St Mary's Island on 23 March, Ramsbottom wrote a note to Youl, telling him that while the vessel was in for repairs 'part of the crew had run away', including his assistant. But he and Captain Hodges had found replacements. Youl was more frustrated the ship was nowhere near a rail line. 'You cannot tell how anxious I have been to get on board with a further supply of ice.'

Three weeks after first leaving London, *Beautiful Star* was finally on its way, with Ramsbottom estimating he had already lost 7500 ova. Success was in balance, and with it Youl's salmon stewardship. He wrote two letters on the same day, but not to the Royal Society, Salmon Commission or the government. One was to an unnamed 'friend' in Launceston, who had it published in the *Examiner*. The other was to 'my dear (William) Archer', who some weeks later passed it to the Salmon Commission and the *Mercury* in Hobart.

Youl made almost identical pleas: 'I hope ... the government will not give up, but make another trial, even should this fail.' And 'I hope the government will make another attempt, should this fail.'

'Great as has been the cost of this trial,' he told Archer, 'even if it should not succeed, it has been money well spent. No one who has not been at sea with salmon ova and tries to keep them quiet and free from motion – which is absolutely necessary – and at the same time have a continuous flow of water can tell the difficulties to be encountered.'

He maintained confidence in his apparatus, declaring it had 'completely overcome' the challenges of managing ova at sea, and that alone 'is worth half the money spent'. He asked Archer to ensure its return. 'I have to beg that the government will ... let me have it all, nothing could act better, and it will be a model for any future attempt.'

In the second letter, he cautioned: 'Many difficulties have already been overcome, but there are one or two more that I know that may prove fatal.' But he remained resolute: 'It will be done, that there is no doubt.'

He wanted recognition of 'this ... as the first real experiment'. The *Columbus* experiment had failed and 'no experience was gained from it', despite Boccius pressing there was no need for running water and that ice was key. And his own *S.Curling* bid offered 'no trustworthy experience' despite Black taking salmon across the line for the first time, and a number of his recommendations being accepted.

Perhaps to dissuade anyone thinking of giving up on him, Youl concluded his letter to Archer by noting he had received 'several applications' to help get ova to New Zealand 'but my wish is to give Tasmania the benefit of the first supply'.

While he awaited news of *Beautiful Star*'s voyage, he experimented with hatching some ova in a soup plate at home. At sea, Ramsbottom was recording in his journal that *Beautiful Star* was 'labouring extremely' in a fresh gale, the ova apparatus swinging 'to and fro with such violence as to render it dangerous to approach it'. Water washed over the deck and seeped into the ova trays. In desperation, Ramsbottom and his new assistant threw blankets over the apparatus and Captain Hodge 'cut out a portion of the ceiling so that water could be ladled out daily'.

Violent rolling made it impossible for Ramsbottom to calculate the precise number of dead ova picked out over four days 'but it could not have been less than 7000'. The next five days brought some calm, but another 2500 ova were dead, and the temperature was rising one degree a day. Ramsbottom could see fewer healthy ova of transparent pink colour, and more with a telltale white spot signifying an ebbing vitality. With the vessel's poor sailing qualities, he 'feared to begin with (ice) before it was absolutely necessary.'

Finally, on 12 April, with the temperature at 54°F (12°C), around the maximum in which ova was said to survive, he felt obliged to employ ice. Two weeks later he placed a letter to Youl onto a passing ship bound for London. Youl was alarmed by his first words: 'We are now very near the equator.' It was 55 days since departure and the vessel had not even reached the equator, and there was not much better news to follow. 'In reference to the ova I have yet some living, though not many, as my loss has been very great.' The ship 'proved in every way against our success ... she rolls and trembles most frightfully.' The apparatus swung so much that ova and gravel was lifted from its bed and tossed about. Dead ova putrefied, and with temperatures still rising he feared 'I should lose the whole'.

Desperate, he 'thought it wise to do away with this (gimbal) apparatus altogether'. He took the twenty good ova left and placed them in the apparatus of suspended trays, which he also reduced in number to cut the distance water had to travel and be kept cool.

'I have some little hopes of landing a few if we can get through the tropics in anything like a reasonable time,' he told Youl, 'but, you see, it is now nearly sixty days since we left London and yet we are barely one-third of our voyage. Had we had a fast-sailing ship I am persuaded that I should have landed some salmon in Tasmania this time.'

He bluntly told Youl that 'should this again be tried', it was 'useless' to think any more of the gimbal apparatus – 'a failure from the beginning' – and the suspended apparatus could not cope with the 'utterly unsuitable character' of *Beautiful Star*. Without a suitable ship such equipment was of 'little use'. But Ramsbottom was unequivocal on the prospects with a suitable ship, changes to the apparatus, and improved water and ice management: 'I could almost lay my life on success.'

His final word to Youl was 'praying I may yet be fortunate'. And he was fortunate, but not as he expected. On the night of 8 May, the melting ice exposed what he had forgotten: the small box of ova and moss loaded as the last-minute, derided experiment.

'On taking up the box,' Ramsbottom journaled, 'found that the lid was broken off, but the ova were still well covered with moss.' He carefully removed the moss, which remained green. Peeling away a portion of the moss 'I had the satisfaction to perceive that, amongst the many dead, there were still some living'. He was amazed. As one of the sceptics not expecting the ova to survive two weeks, he had 'no expectation of finding live ova ... even had the box been perfect', which it was not after pitching about. After removing about 250 dead ova 'with all the care and speed possible' he had 'nineteen living, to all appearance in good health'.

Ramsbottom sensed a light-bulb moment. 'This little experiment will no doubt prove of much future value, as indicating a new and successful method of transporting salmon ova to distant countries.'

After utilising the final three ice blocks, the last of the primary ova load died 30 minutes later, at a temperature of 59°F (15°C). But the moss box ova lived eight hours longer, even as the temperature rose to 65°F (18°C). They had survived 74 days since leaving London, 88 days from being taken from their parent fish. Ramsbottom was amazed: 'I can only add ... my extreme astonishment at the ova surviving so long under such tremendous disadvantage.' And the ova had survived to 22° south, salmon's deepest foray into a new hemisphere.

There was now nothing for Ramsbottom to do but wait patiently for Captain Hodges to get *Beautiful Star* around Cape Hope and head eastward to an anxious and impatient Hobart, where mail deliveries had progressively catalogued the selection and journey of *Beautiful Star*, numerous delays, and an appearance in a list of shipping 'casualties'.

The *Mercury* said the troubles 'almost preclude the hope' the voyage would be successful. And if this proved true, it believed the government's intention was not to repeat it under similar conditions or spend more public money funding experiments run by 'no one [with] a direct and positive interest in the success of the experiment'.

The fate of the arriving ova was to be indicated by the vessel hoisting the flag of London above the Union Jack if it was a live delivery, or the London flag over the Liverpool flag for death. But instead of the usual relay of flag signals up Storm Bay to Hobart, the Master Warden wanted to employ a new telegraph system which the *Mercury* proclaimed sent messages 'along the electric wire with the speed of lightning' in another triumph of science over distance.

After a 142-day voyage the suspense was over, the electric message succinct: 'Arrived 24 July, three-masted schooner *Beautiful Star*, from London – ova all dead'.

News of another failure raced around the colony and divided it. In Youl's hometown, the *Launceston Examiner* said it was disheartening but maintained the preparations were 'proper', and hoped 'neither those at home nor those here will accept this last defeat as final'. In the rival south, Youl's eschewing of instructions, tardy communications, and criticism of 'ignorance' and 'apathy', had not endeared him. The new Salmon Commission said the London Colonies Association had many members of 'great intelligence' with a deep interest in salmon 'but its chief direction fell into the hands of our fellow colonist, James A. Youl Esq.' At a Royal Society meeting, commissioner Morton Allport did not hold back: the *Beautiful Star* choice was 'fatal', consigning the ova to 'utter destruction'. It would have been far better, he said, 'to have delayed the experiment till a suitable vessel could be found, even if we had waited five years'.

A few days later, Allport and fellow salmon commissioners prepared a report to the governor. As gentlemen, they acknowledged 'the zeal and energy displayed by Mr Youl in the performance of this voluntary and patriotic duty' which deserved the highest commendation. But the outcome was 'disastrous'.

'Mr Youl was, from the first, fully impressed with the importance of embarking the ova in a vessel at once swift and roomy, and sailing direct to the Port of Hobart Town ... essential, in the opinion of the commissioners, to the success of the undertaking.' But his 'fatal error of judgment' in selecting *Beautiful Star* for a cheaper price 'rendered success all but impossible', regardless of the apparatus and ice arrangements.

Despite the deep disappointment, no one wanted to abandon the quest. Commission Chairman Robert Officer trusted that the 'unalterable motto' would be 'try again' until complete success was achieved.

But if there was to be another attempt, the big questions were 'by whom?' and 'at what cost?'.

The commissioners were impressed with Ramsbottom, who appears 'to be thoroughly acquainted with the duty he undertook to perform, and to have discharged it with much diligence and zeal. They believe that if another effort of the same nature should be made, it could not be confided to a more efficient agent, or one more likely to ensure a successful result.'

The commissioners were heartened by his conviction that the right management and apparatus 'would almost certainly render the introduction of salmon into Tasmania a certainty'. As for his surprise revelation of ova surviving longer in a moss box embedded in ice, this was worthy of 'further exploration'.

The shipment and salmon ponds preparation costs had run to £2319 (£240,000 today, A$500,000). Youl wanted the apparatus returned for reuse to help keep the cost of the next voyage below the estimated £1000. But Ramsbottom urged a doubling of ice to 50 tons (45 tonnes), which meant additional freight charges and a larger icehouse, which he wanted lined with expensive tin to retain water purity. He estimated £1890 (£196,000 today, A$408,000).

The commissioners supported a higher cost to make the arrangements 'as perfect as possible and provide every safeguard ... against future failure'. It was 'a very small amount' compared with the benefits to be conferred on Tasmania and any neighbouring colonies willing to contribute. The government agreed to £1725, with Southland in New Zealand committing £200 and Victoria £500.

The Salmon Commission wanted Ramsbottom back in England 'with the least possible delay' to make 'the arrangements and preparations for the renewal of the experiment', then return to manage salmon breeding at the new ponds. The government agreed to his £60 return passage to London and a salary for two years at £156 a year. Making no mention of Youl, the commission said Ramsbottom would 'put to the test' more experiments of the preservation of ova in moss. He was hurried to Melbourne to board *Great Britain* for London and 'await the instructions ... as to the renewal of the experiment'.

Seven months after *Beautiful Star*'s departure, Youl and the rest of Britain learned of its failure. The *Argus*' despatch to London was as supportive as possible, not specifically mentioning Youl by name. And it maintained the experiment 'has proved ... by the length of time the ova lived that, and the novel manner in which the vitality of the other ova was preserved, that salmon may be introduced into Australian waters if the experiment be made on board such a fast ship as the *Great Britain*, which can be relied upon to make a fast passage.' Especially as it had again just reached Melbourne in 60 days, less than the time ova had been kept alive on *Beautiful Star*.

Edward Wilson felt it 'mortifying' that success was thwarted by not having a suitable vessel. Any government except England with such maritime power would readily facilitate such acclimatisation efforts: 'We might as well expect assistance from the moon as from the Admiralty.'

Whether the Tasmanian Government would continue utilising the Colonies Association, knowing it would undoubtedly defer management to Youl, was now in the hands of a new governor, Colonel Thomas Browne, warmly welcomed as representative of a 'new order' as first appointment after the colony gained responsible government.

The *Examiner* argued that 'to Mr Youl and the gentlemen connected with him (that) the colonists are much indebted for the time, trouble and outlay they have so willingly accorded for their benefit'. In the political capital, the *Mercury* said that because of the 'absurd blunder' to despatch ova on such an 'absurdly small vessel ... like a waif upon the waters' the attempt was 'beaten by the mismanagement of the persons in charge of the experiment at home ... most unfortunate it is that the experiment should have been conducted by those ... with so little judgment.' Both Ramsbottom and Alexander Black had laboured under causes and mismanagement beyond their control, and 'so little judgment'.

Youl would never forget the criticism and derision. Like most in the north of the colony, he had little time for elites 'down south' – and had never forgotten their ingratitude for his father's pioneering efforts, nor what he saw as insufficient support, gratitude or recognition of his voluntary efforts to progress salmon, female emmigration, financial aid and mail services.

'When the news of the (*Beautiful Star*) failure ... became known in Tasmania, every kind of abuse was heaped upon me by the Press and the people,' he later related, his reputation readily slighted despite giving 'the whole of my time and services gratis'.

A London friend, Arthur Nichols, who wrote articles on the salmon quest, said the colonists had placed their funds and confidence in the hands of one who 'held out no extravagant promises'. He had merely undertaken 'to do his best towards the solution of a problem of which no man then held the key'. His plans and spending were 'carefully considered and well warranted', but the colonists 'stultified themselves completely by their censure, as unjust as it was ungenerous and illogical'.

Youl was angered, Nichols later wrote, by those who felt the quest 'would never succeed so long as the experiment was entrusted to such incompetent persons as Mr Youl'. They also 'tried to get someone else' and considered the merits of calling tenders or offering a larger reward to encourage others.

But it was finally concluded in Hobart that any changes would cause further delay and risks, and the causes of failure were now so well known that 'success could hardly fail to crown another effort' except in the case of a shipwreck or similar disaster. Directing another voyage from Hobart was impossible, so there was no choice but for 'all direction to remain with the Australian Association'.

The Salmon Commission understood the association would 'practically devolve on Mr Youl' but clearly wished it would not devolve too much. It told the government that for the 'safe transport to the Antipodes', Ramsbottom, strongly endorsed by Allport, was 'now making the necessary preparations at home'.

In a letter to Youl, the commission politely affirmed 'a lively sense of (his) zeal and energy ... earnestly requesting his continued co-operation on their renewed endeavour to effect this great object'.

But Youl was never going to cede direction, or honour, to anyone else. This was his mission.

11

THE ICEMAN COMETH

Robert Ramsbottom was 'astonished' when he heard of his son's *Beautiful Star* experience. It challenged his conviction that it was 'impossible' for ova to travel and survive across the world, but did not change it. Whether the combination of moss and ice in a box was a solution to keep ova alive all the way to Hobart, then successfully hatch, remained uncertain.

Clues had long existed but never been pursued or connected: in the 1700s with German Stephan Jacobi and Swede Charles Lund with their hatching boxes and use of foliage; in the 1840s, with Frenchman Joseph Remy's perforated tin boxes, and former Youl neighbour John Wedge's use of damp moss in a glass bottle; in the 1850s, Gottlieb Boccius' declaration that flowing water was not essential; botanist John Bidwell's conviction that ova could be shipped in moss packed in ice; and Andrew Young's observations of hatching salmon ova in frozen baskets of gravel. Other elements were evident in the experiences of Swiss

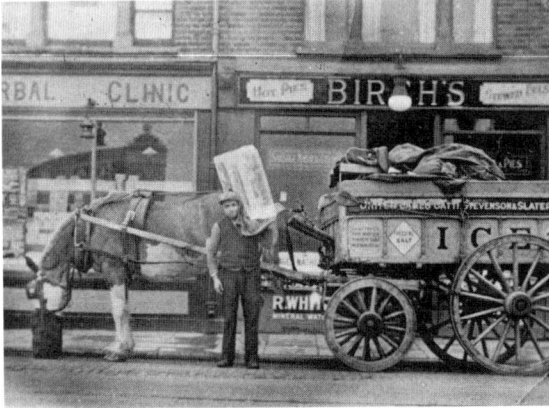

Early London ice delivery. Photograph, courtesy Friends of Islington Museum.

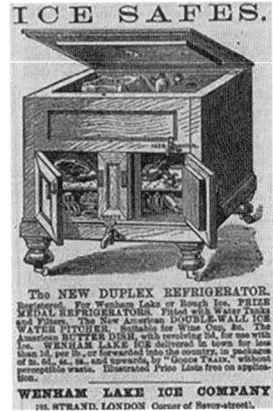

Illustration of ice box

Professors Louis Agassiz and Karl Vogt; Englishmen Dr John Davy and Professor John Pepper; Irishmen Thomas Brady and William Ffennell; and Frenchmen Professor Victor Coste and Zepherin Gerbe with their ova enclosed within moist moss or cloth.

And in the streets of London, Wenham Lake Ice Company men were conspicuously loading wheelbarrows to deliver ice blocks to those with new 'American' or 'polar refrigerators', the new technology for preserving meat, fish, game, poultry, milk, butter, ice cream, jellies and wine.

When the wooden box containing ova was first placed amid ice blocks on *Beautiful Star*, no one, other than the Wenham Lake Ice manager, Henry Moscrop, had any faith in it, including Youl, who in an initial report to the Colonies Association after the vessel's departure admitted to having 'almost forgotten' to even mention the box. His builder, Thomas Johnson, also said there was 'so little faith' that everyone present 'declared all the ova would be dead before the vessel got to Gravesend'. Ramsbottom Sr said he would not be a party to such a 'ridiculous experiment' and newspapers reported 'scientific authorities ... ridiculed the idea'.

But after the unexpected outcome, Youl's position was transformed as he sensed he might be close to unlocking the key to a miracle. Now, he said, 'all at once, as if by inspiration, I thought of the packing in moss ... I hit upon the idea of a small deal (pine) box. When I produced my little box ... all those present on board the ship ridiculed the plan as the most absurd. Fish, they said, must be hatched in water, and not in moss.' He had been 'obliged with his own hands to pack the box and almost furtively deposit it in the icehouse of the *Beautiful Star*'.

The self-serving take must have irked others, including Coste, Gerbe, Brady and Moscrop, but an emboldened Youl mused that if some experiments proved successful, he might invest in a private voyage to deliver the holy grail without having to rely on government authorities and endure what he called 'further obloquy'.[1]

'So impressed am I with the little experiment in the box with moss that I mean to try an experiment at my own expense this year,' he told the Salmon Commission in October 1862.

Youl just wanted to completely satisfy himself that a continuous stream of water was not essential, and that deprivation of air or light would not be fatal. Only then would he give up any complex imitation of nature. Coste had advised in Paris 'the experiment of sending the ova in moss, protected and kept cool, very well worth trying', suggesting that as each degree or two of lower temperature substantially increased retardation, 35°F (1.6°C), just above freezing, hatching could be delayed for 95 days. But his associate Gerbe had cautioned about such a long voyage.

Ramsbottom Sr said he and William would collect the ova for Youl to test whether he could send 20 small boxes of 300–500 ova in an icehouse all the way to Hobart. But the veteran fisherman still thought Youl's vision was madness, wishing he 'understood spawn as well as I do'.

'I want you to remember what I am going to say about sending spawn in moss and ice, that it will not answer. If you stop the process of hatching, you kill the spawn. You may retard it but you cannot stop it ... when you have handled spawn as long as I have done you will find it tedious kind of stuff.

'I don't think it is possible to transport it to Australia. We can send young fish but I doubt the spawn in any way. You can send young fish, but that is all. I am confident spawn cannot be taken. You might as well try to fetch Australia to England.'

11. 1   Disgrace from condemnation

Youl was stirred but not shaken, remaining determined to land salmon on what he had decided would be his final attempt. And he intended doing it his way, with Nature's 'profusion' of ova. Mistakes had been made, but as Jules Verne was writing in *Journey to the Centre of the Earth*, 'science ... is made up of mistakes, but they are mistakes which it is useful to make, because they lead little by little to the truth'.

Youl fretted his mistakes could lead to others besting his quest. Past failures were instructive, as the *Argus* put it, for 'such individuals as may be ambitious of winning a name in the pages of Australian history by the introduction of salmon into the Southern Hemisphere'. Whoever achieved it would be entitled to 'a very handsome bonus' from all the colonies.

More were being drawn to what naturalist Frank Buckland described as 'gold nuggets under our noses ... (we had just) not stooped to pick them up ... to be converted to bank notes.'

New acclimatisation societies in New Zealand were also beginning to actively pursue the prospects of transporting salmon ova from Columbia or Oregon rivers, seeing a six- to eight-week voyage as cheaper and less risky. A New Zealand ship captain proposed transporting eggs from the Sacramento River, California, in four-ton boxes of ice. A Birmingham farmer and fish culture hobbyist, Andrew Johnson, planned to migrate to Canterbury with a tank of young fresh-water fish, including 800 salmon and 600 trout.

Fish propagation was also being developed in North America, with Institute of Agriculture members being told 'trout can be artificially produced with greater profit to the producer than any kind of meat'. As the baton was passing from Johnny Appleseed[2] to Johnny Fishseed, Youl might have noted a surprising connection. William Ramsbottom's younger brother, James, had migrated to America and placed an advertisement in the New York *Herald*: 'Mr J. Ramsbottom, son of the well-known and successful pisciculturist, Mr Robert Ramsbottom of Clitheroe, England, will shortly be open to engagements. For further particulars address JR, Herald Office, NY.' A New York *Tribune* series

---

11. 2   John Chapman, better known as Johnny Appleseed, was a pioneer missionary-nurseryman who introduced apple trees to large parts of America; he attached symbolic importance to apples, just as symbolic importance was attached to salmon.

on transport and hatching methods soon cited 'professor Ramsbottom', whose brother 'is now engaged at propagating trout in Australia', as available for hire for artificial propagation of salmon and trout.

The superintendent of fisheries in Quebec, Richard Nettle, had also 'long watched' and admired the Tasmanian salmon efforts. 'But it had long been a surprise to me that the nearest place from whence salmon ova could have been obtained were not sought for.' He had a proposition to benefit 'all places that may be adapted to the habits of the salmon species within the Southern Hemisphere'. He put to Sir William Denison, who had supported the salmon quest as Governor of Van Diemen's Land in 1852 and was now Governor-General of all the colonies, a plan to rail spawn from tributaries of the St Lawrence to New York in three days, then by mail steamer to the Isthmus of Panama in two days, and then across the Pacific to Australia in 30 days.

With just 48-hours' notice, wealthy Canadian American shipping owner, Roderick Cameron, whose company sent ice to Melbourne, jumped on board. He would 'feel proud indeed' if he could introduce live spawn across the Pacific to the Antipodes. After delaying the sailing of his *Montreal* from Quebec so Nettle could secure ova, he was confident they 'will all arrive live!' But 'if the first attempt fails, a second shall not'.

This was a pivotal moment in the Antipodean quest for salmon. On Anniversary Day, when the colonies celebrated the 26 January 1788 proclamation of British sovereignty, the *Argus* declared: 'That we shall have salmon introduced to our waters is only a matter of time. It becomes a race as to whether the British fish, or the salmon of the St Lawrence, shall be first placed in Australasian waters.'

The first North American attempt, the fourth bid to land salmon in Australia, proved an 'utter failure' – all ova dead within two days after little preparation and ice not arriving in time – but Cameron vowed to try again. *Field* said if anyone proved the success of icy suspension of hatching on a journey across the Pacific or Atlantic then 'transporting fish from any part of the world to another will ... be at an end'.

Having an international race was not what Youl wanted. He had been in the quest the longest, and after enduring a decade of failure, expense, ridicule, and 'abuse', he was running out of patience and time. Much was hanging on his experiments – designed to unambiguously prove whether moss-ice was the solution – at the Wenham Lake Ice Company office in the Strand and its huge ice vaults at Blackfriars. In January 1863, the Ramsbottoms delivered 5000 ova from the Ribble in their hand-fashioned ova transportation device: a wine or soda bottle, in the bottom of which they made a small hole 'the size of a quill' around which sealing wax encased a wooden peg. Fresh water passed into the open neck of the bottle, over the ova and, with the peg withdrawn, out. Ramsbottom Sr had once dropped a bottle and not a single egg hatched, which was why he thought the notion of transporting ova across rough oceans was fanciful. Hearing the tale, Youl's author friend Arthur Nichols was also convinced 'the task of conveying them to Australia was hopeless'.

In the Strand office, Youl first had ova embedded in layers of saturated moss in a box inside 'an ordinary meat safe' with a block of ice in a separate compartment but without water. In another section of the meat safe he put ova in a dish of moss and a small amount of water. In a large icebox, ova were packed in wet moss inside a box with perforated holes, with pieces of ice constantly kept on top of the box so melted cold water passed through the moss and ova and out of holes at the bottom. Youl satisfied himself the box was air-tight by enclosing a lighted candle, which went out soon after the lid was closed. Temperatures in the meat safe and icebox were aimed to be about 36°F (2.2°C) – just above freezing.

Youl then headed to the company's five massive ice vaults under the arches supporting Waterloo Road. Forty-feet high and 70-feet long (12 by 21 metres), the vaults housed 2000 tons (1814 tonnes) of ice amid huge mounds of sawdust, a non-conducting medium to minimise thawing. Youl placed between 200 and 600 ova in eight boxes of damp moss buried amidst ice blocks two-feet thick.

'In these spacious dungeons, in silence and in darkness, old King Frost is cooped a close prisoner,' the aptly named author Andrew Wynter described. Youl heard 'many persons' advise he was burying ova in 'icy graves', the moss merely a 'shroud'.

While waiting six weeks before his first ova inspection, Youl was drawn to *Field*'s office window in the Strand. Frank Buckland had developed a 'rude but effective' hatching-apparatus model to provide 'pleasure and amusement (for) many thousands of people who have certainly never seen a salmon alive before' and promote the scientific importance of fish propagation. Angling editor Francis Francis was enthralled that this 'delicate … little stranger … helpless little thing' could become a magnificent 20-pounder (nine kilograms) giving an angler 'a full half-hour's hard work and awful excitement'.

Newspapers reported 'wonders will never cease' and urged readers to see the shopfront miracle for themselves. Crowds blocked the roadway, and the *Newcastle Journal* marvelled: 'What shall we come to next?' Salmon had been hatched in an office window in the heart of London, and in the same street ice shipped across the Atlantic was 'freezing English fish for transport to Australian rivers'. The *Journal* saw 'shades of Columbus and Captain Cook!'

Youl invited Buckland to witness his ice experiments. Buckland was doubtful but the quest was 'exciting great public interest', and Youl had supplied ova for the Thames Preservation Society's efforts to replenish 'Old Father Thames' between Hampton and Teddington. Leading scientists did not share what the *Sun* called 'these ardent Waltonians' pleasing faith in the possibility of replacing the lost breed of fish in the chief river of England'. Dr Albert Gunther, author of an eight-volume *Catalogue of Fishes* in the British Museum, said any salmon trying to migrate back from the sea to its native river 'would turn back in disgust at … London Bridge'.

Buckland acknowledged 'we are laughed at for our pains'. But to him the quest to send salmon across the world was 'THE question of the day … there are many persons who tell us that salmon can never be taken to Australia, and if they ever arrive there can never live there. Anyhow we shall see what is to be done. If it is a failure, then it is a failure. But if a success, what a success!'

It all hinged on the experiments. After 45 days, Youl unscrewed boxes in the meat safe and icebox and was pleased to see the ova remained healthy. When the eggs were still in 'perfect health' after 70 days, Buckland declared the initial results 'decidedly … encouraging …

[and] everybody will be glad he [Youl] intends to persevere, for the key to the secret must be somewhere about, and we mean one of these days to find it.'

Then it was time for the first check of the boxes entombed in the ice vaults. Youl's mind was fevered. Many had predicted the ova would be long dead. Was that true, or was he close to unlocking a secret?

Holding his breath, he unscrewed and gently lifted the first box lid. Taking a large magnifying glass to look inside, Youl felt an 'indescribable pleasure ... our little friends [were] alive and perfectly healthy!' He was exhilarated and relieved. 'All right!' an excited Buckland exclaimed. 'All right!' After much backslapping, Buckland carried some away 'in triumph' to the Zoological Gardens to 'ascertain for Mr Youl and our Australian friends whether they would hatch out after such a long period of freezing'.

To Francis Francis, this was the crucial question, 'not whether by extra cold [one can] retard the development of the egg, but for how long can we do so safely? And then how we are to get a sufficient supply of ice out to Australia? It may be the eggs will eventually hatch but what sort of embryo is the result? That is the point. Will they be capable of rearing? I fear not.'

Francis wanted more than opinions, which he lamented were never in short supply, with 'those who know the least about a subject the most ready to impart their views'.

The ova had survived 10 weeks, about the time it took the fastest steam ships to reach Melbourne. But Youl wanted to know if eggs could survive at least 90 days, the likely time for a sailing vessel to reach Hobart, and what he saw as the 'highly dangerous' time limit before ova's immersion into natural streams.

Waiting for further inspection of the ova, Buckland praised Youl for his 'pain in bringing these experiments to a successful issue' but he was the one seeking and drawing attention. He took his fish-hatching apparatus model to the Royal Institution, Royal Society of London, and Royal Microscopical Society, and showed it to the Prince of Wales at the international dog show at Islington. In a newly released *Fish Hatching* book he said: 'The reader must give me credit, just as he gives credit to the astronomer who will tell him what is on the other side of the moon, a planet which has never shown the reverse of her lustrous orb to mortal eyes.'

He even made himself a salmon. Wanting to know what a migrating salmon faced at a weir ladder, he astonished bailiffs by jumping into the Corrib in Galway. 'Oh! You shining lovely creatures,' he wrote. 'Whence come ye? Whither are ye going? Oh! that I had scales and fins for five minutes. I imagined myself a salmon.'

When a Zoological Society staff member mistakenly tossed into the street a basin of tiny salmon which Buckland had intended to show an audience, he rushed outside with candles, spoons, and camelhair to 'save the grasping little wretches one by one'. When a crowd asked what he and helpers were doing, Buckland answered 'catching salmon'. To which one man gasped 'Catching salmon! In Hanover Square! Good gracious, what next?' before rushing away, thinking he had come upon 'a party of midnight lunatics'.

But 'lunacy' was giving way to truth. On what Youl saw as the pivotal 90th day, ova embedded in a box of moss remained vital. He was now persuaded that neither water nor full light was necessary, and moss was a viable security blanket. All the toil and trouble, expense and complexities of water tanks, piping, suspended trays, gimbals and running water 'would be finally abandoned ... I have matured a plan of my own for placing the boxes with the ova in the icehouse'.

After 120 days, four more boxes in the ice vaults were opened. The result was more remarkable and memorable. 'In every box we found the ova alive', Youl rejoiced, judging them 'as perfect as in their natural spawning beds, or in the best constructed artificial breeding ponds'.

120 days! That was long enough for even the slowest sailing voyage to Hobart. 'The experiment has proved more successful than my most sanguine expectations had led me to believe was possible,' Youl told friends. 'I have little doubt that by this means by this time next year salmon fry will be swimming in the artificial breeding ponds prepared for the ova in Tasmania and New Zealand.'

After 144 days the last of the vault boxes was opened. Still the ova were vital. Even a box of 146-day old eggs, dispatched to the Ramsbottoms'

Clitheroe home on a two-day journey 'by ordinary goods train' and cab, arrived in excellent condition and hatched. William Ramsbottom declared he and his father were 'living witnesses of the success of this grand experiment'.

The solution had been found by utilising ancient elements of moss, wood, and ice. Nature's truth was that it was not about 'freezing' the ova, as many had called for, rather that substances above freezing and buried in a box surrounded by ice would remain at a low temperature but not fall to the fatal freezing point.

It was a remarkable and almost accidental breakthrough. The solution had been suggested by others a decade earlier but been overlooked or ignored. Then it emerged through a trial box which Youl himself almost forgot to tell colleagues about, and which according to him Ramsbottom had also forgotten about it and only came across it 'by getting a thump on the head from it as a reminder of it being there', while Johnson similarly told the British Association Ramsbottom 'stumbled over the box which had been forgotten'.

But who would be remembered for the breakthrough? With the rising prospect of success, and understanding failures were orphans while any success had many fathers, Youl disingenuously asserted: 'There is no doubt with every disinterested person who took an interest ... that it was entirely owing to my former experience in the shipping of ova and building an icehouse on the *S.Curling* that I was enabled to work out M. Gerbe's suggestions ... to obtain success.' This, despite having previously said 'no trustworthy experience' came from the *S.Curling* voyage, and his initial rejection of French advice.

He complained that the Salmon Commission and Royal Society saw merit in the moss box but 'did not mention this in their correspondence with me ... [nor] give me any credit for having put the little box on board as an experiment'.

To Youl's annoyance, the question of credit became vexed. Buckland acknowledged the 'energetic superintendence of Mr Youl' to the Royal Institution but put himself into the frame by saying 'after a great deal of trying we have found out how to do it ... [through] my experiments'. The Acclimatisation Society of Great Britain, which Youl had just joined, also said Buckland had solved the problem of how to transport salmon to Australia, having 'discovered that salmon ova might be frozen'. The *London Review of Politics, Society, Literature, Art and Science* lauded the labour of 'Dr Buckland and his friends', saying that when England left everything to individual rather than government effort, 'much more praise is due the amateur efforts like Mr Buckland's than is usually accorded ... who ever heard until lately of a water farmer!' Buckland's co-founder of *Land and Water*, William Ffennell, the first inspector of fisheries for England and Wales, praised the 'enterprising zeal, perseverance and scientific skill' involved in experiments 'under the direction of Mr Frank Buckland', adding that the names of James Youl and Edward Wilson 'should be recorded as taking a foremost part, undaunted by every difficulty'.

Then Wenham Lake Ice's Henry Moscrop claimed the credit in letters to the London *Review* and the *Times*. He had suggested putting the box of ova within blocks of ice on *Beautiful Star*, and after learning of the ova lasting 74 days, 'I was satisfied that had this ice lasted long enough the ova would have reached the destination in good condition ... I offered to put the matter to the test at my own expense and to ascertain how long the ova could be preserved in ice.' His version had support from Youl's builder Johnson, who had become deeply engaged in the salmon project, and in a paper read at the British Association, acknowledged 'the value of a suggestion by Mr Moscrop, viz, burying the ova amongst ice ... this important discovery has [now] been fully tested ... placed beyond any doubt.' In a footnote, *Review* founder-editor Charles Mackay said the claim was 'just' and 'Mr Moscrop's practical experiments are the first that have ever been made to prove a very important point ... meritorious.'

In Tasmania, Morton Allport promoted his Royal Society as having first suggested using ice, and it was his 'duty' to secure public credit for William Ramsbottom realising the 'entirely original' idea of ova in moss and ice. For its part, the Salmon Commission claimed it had instigated the experiments which led to 'one of the most valuable discoveries ever yet made in the art of pisciculture and must ever indicate an important era in its history'. Youl's former Launceston Horticultural Society colleague, Dr Matthias Gaunt, said his ice advocacy had been dismissed by 'great men puffed up with their own conceit ... Mr Youl's mind has been so led away in trying complicated and expensive, but useless, experiments.'

It irked Youl to be put in the shadow of others who had done much less in a practical sense. He had good cause to be determined that his next bid would definitely be final and conclusive, delivering due and eternal acknowledgement for what the *Cornwall Chronicle* had suggested long ago: the Mr So and So who introduced salmon into the rivers of 'the Britain of Australia'.

12

A BOLD GAMBLE

Youl was not one for reporting to anyone, but 'in compliance with your request to be informed' he obliged the Salmon Commission with the results of the ice vaults experiments, and advised he would ship ova in the same manner, and perhaps some in 'a small freezing machine such as ice creams are made with', as Dr John Davy and Thomas Ashworth suggested.

Commission chairman Dr Robert Officer quickly boasted to the government that the 'most valuable discoveries' seemed to remove all doubt about conveying salmon ova to the colony. But Youl still had to resolve the issue which cruelled his *S. Curling* and *Beautiful Star* attempts. Finding a fast, first-class ship sailing direct to Hobart, at the right time, and for the right price, was 'the greatest difficulty I have had to success'.

Edward Wilson again pressed for Her Majesty's Navy to help create salmon history, especially as the Navy became formidable after the discovery of the rich fishing grounds of Newfoundland. 'One ship, in a single trip, might convey to a new country some great thing which might be the making of that country.'

The proposition startled the Secretary of State for the Colonies, the Duke of Newcastle, who deferred to the Admiralty. Wilson rated it 'one of the most impracticable departments of an impracticable government', but he took up the fight with its First Lord. The acclimatisation of fish and other species to the Empire's 50 colonies, he wrote, had the potential to add 'wealth and happiness and multiply the imports and exports of the Empire'. But despite the 'great interest' of Queen Victoria and the late Prince Consort, the Imperial Government had 'done nothing directly to give material assistance'.

Years of 'disheartening and expensive losses' were mainly attributable to the want of suitable vessels, when the Navy could help deliver results 'which can never be arrived at under the present imperfect system'.

The application to use noble man-of-war vessels to transport fish was probably received with 'astonishment and disgust', one newspaper reported, but the First Lord of the Admiralty, the Duke of Somerset, agreed to at least receive a delegation.

Oh no, he quickly said, it was impossible to detach Navy vessels for such purposes. It was 'usual in this country to leave such operations to the enterprise of private individuals'. This was the very point of frustration, the delegation argued. The rival French Government's fulsome support of acclimatisation projects throughout its empire meant pisciculturist Victor Coste had 'one or two government steamers always at his disposal'. This meant the French had left Britain behind. And there were precedents: Her Majesty's Navy had supported scientific missions such as those of Sir Joseph Banks, the HMS *Blonde* to South America and Hawaii, and the pursuit of the Arctic's north-west passage.

The Duke described these as 'isolated' operations, and he could do no more than 'consider' the request. Four days later the Admiralty secretary advised 'my Lords Commissioners of the Admiralty ... regret that it is not in their lordships' power to appropriate any of Her Majesty's ships of war to that object.'

After an ensuing paper war about whether providing even the least valuable Navy vessel was incompatible with advancing empire interests, the Admiralty reluctantly messaged all commanders that as long as public service requirements were followed, and no expense was to be incurred, they had permission to assist any accredited agency of the acclimatisation society 'who may apply to them through any of her Majesty's Foreign Ministers or Consuls, or through the Governor of any of her Majesty's Colonies, for the transport of specimens'.

Field, whose owner John Crockford had been part of the lobbying, declared 'an epoch in the history of acclimatisation', but while Youl initially thought 'there is some hope' he realised there was little prospect of quickly securing any willing commander, let alone a Naval vessel sailing direct to Hobart.

With the spawn season fast approaching, he would have to quickly find his own vessel quickly, and a better option than his fatal *Beautiful Star* choice. Several plans were again aborted when owners had last-minute fears of melting ice or could not finalise a departure date. Then Youl spotted an advertisement of Money Wigram and Sons, one of the first London ship owners to transfer their trade vessels from India to Australia to cash in on the goldrush years. In East India Dock, they had what the *Morning Chronicle* described as 'a celebrated clipper', the *Norfolk*, leaving for Australia on 20 January via Plymouth to pick up passengers trained from Paddington.

Sheathed with red copper, the 953-ton (865-tonne) three-masted, five-decked, 196-foot (60-metre) vessel was impressive, known for delivering millions of pounds of Australian gold and wool. Its departure date aligned with Youl's schedule and was promoted as 'one of the fastest ships in the Australian trade', making runs to Melbourne in 67, 68 and 70 days. But that was Youl's dilemma: both the Tasmanian Government and the Salmon Commission had again expressly called for a fast, direct sailing to Hobart. If he chose *Norfolk*, and the ova arrived safely in Melbourne, the Victorian and Tasmanian governments would have to arrange and fund a transfer for the final leg.

Norfolk. Photograph, State Library of South Australia, PRG 173/19/63.

Youl weighed his options. He was hopeful but could not be fully certain his ice-moss experiment would succeed as well at sea as on land, but felt *Norfolk* was the best chance yet to win the salmon quest. And he had decided it would be his last attempt. If it failed, the authorities would not turn to him again. He confided to Wilson: 'I shall give it up.'

He made his decision, a bold gamble: getting ova quickly to Melbourne on *Norfolk* outweighed any risks of a transfer to Hobart or waiting in hope of a better arrangement.

Or being beaten by someone else.

There was no time to inform Hobart. By the time any correspondence arrived in Tasmania the vessel would already be on its way, and as for any approval he believed he had 'supreme authority' from the Colonies' Association.

The *Norfolk* name also seemed propitious: Van Diemen's Land was first circumnavigated 70 years before by Matthew Flinders in a Norfolk Island-built sloop called *Norfolk*, and the salmon ova was destined for ponds near New Norfolk.

The salmon quest had caught the eye of the shipping firm's chair, Charles Wigram. From its Blackwall yard it had a fleet of 30 'blackwallers' named after counties such as Norfolk, Kent, Sussex, Yorkshire, Essex and Lincolnshire. On Boxing Day 1863 Youl received a gift beyond his expectations: Wigram 'nobly [has] given me room for the salmon ova in the best and fastest of their ships, the *Norfolk*, to sail on 20 January. The greatest difficulty has been overcome.'

When the reality of the chairman's generosity became clear – a sizeable icehouse and 30 tons (27 tonnes) of ice would consume 50-tons (45-tonnes) worth of valuable cargo space – Youl worried the offer could be withdrawn. He quickly offered 100 guineas (£1066 today, A$2200) 'out of his own pocket' as compensation for the loss of cargo space. The company resolved it would honour the offer, but only on the condition Tasmania and Victoria understood it was 'the importance it is to the colony to introduce the salmon into its waters that induced us to accede to give you the necessary room', and that the generosity would not be overlooked in future shipping arrangements.

But even the finest ship did not guarantee safety at sea. Beyond rough seas and icebergs, ships of war threatened. On *Norfolk*'s last voyage to Melbourne a Confederate Navy war sloop, *Georgia*, staged an armed boarding off the island of Trinidad. After initial panic, the *Norfolk*'s captain, his passengers and sailors exchanged newspapers, and the news that Civil War leader, Stonewall Jackson, had died after being accidentally shot by one of his own men.

Youl's decision to select a vessel sailing only to Melbourne astonished Hobart, with Salmon Commission chairman Dr Officer 'startled'. But upsetting authorities in Hobart had never worried Youl, and everything seemed right: *Norfolk*, the moss box and icehouse, Robert Ramsbottom pursuing ova, and William Ramsbottom as midwife. But on the cusp of what he hoped to be a triumphant end to his long quest, an alarming letter dropped at his Clapham Park home on 5 January 1864, 15 days before *Norfolk*'s departure. 'To my great dismay', Ramsbottom Sr. wrote of his despair that every salmon caught in the Ribble had already spawned. Ramsbottom asked: 'What am I to do?'

No ova! Youl could not believe it. Everything was ready for a defining chapter in his life and the history of fish and Tasmania, except for the fish. Surely, salmon would not let him down at the final hour? He told Ramsbottom he and William must get to the Dovey, the Welsh river which as late as February had produced ova for the *Beautiful Star* attempt. Another son, Westall, went to the Ribble, and Youl assigned Thomas Johnson and his brother to pursue the Tyne, and later the Ettrick and Tweed.

In desperation, Youl made a national appeal through the *Times*. 'I have just now most unexpectedly met with a great difficulty in my endeavours to carry out a national undertaking, in which I have no pecuniary interest.' He was 'so pressed for time' that only through the

Times could he plead with proprietors on the Dovey and Tyne to afford the Ramsbottoms' and Johnsons' every assistance.

He also urged 'any of your readers' to provide any unspawned salmon 'so that I may not lose the only chance I ever had of fairly trying to get this noble fish to Australia'. The appeal captivated the nation. Along with the first Football Association matches, this was an affair the public could warm to against darker issues of the day: the conspicuous absence of a mourning Queen Victoria stirring republican sentiment; the American Civil War; New Zealand Maori wars; and the Danish-Prussian dispute heading towards a crisis.

On 12 January Youl was beside himself. 'I nearly despaired of success ... no ova ... obtained for me [and] the ship positively to leave the docks on the morning of the 20th.'

Ice was his ally, but King Frost his enemy. With temperatures down to eight degrees below freezing, thousands enjoyed sliding and skating in London's parks, but numerous rivers iced over, obstructing navigation of small craft and thwarting ova collection.

With time ticking, Youl rushed to Worcester to secure access to net the Severn and one of its tributaries, the Teme. He called for anyone who could secure 20,000 ova from anywhere, be it the Dovey, Tyne, Tweed, Ettrick, Ribble, Teme or Severn, to 'come by first train to London'.

Six days before *Norfolk*'s scheduled departure, a freezing, despondent Ramsbottom Sr had still not found suitable spawning fish. He wrote to Youl, 'begging that he might leave, as he was only wasting time and money to no purpose, as it was too late in the season to get any ova'.

Then came an unexpected turn when Sackville Phelps, secretary of the Dovey Club, stepped in. He usually defended the privileged entitlement of wealthy esquires with riverside properties to fish without being 'mobbed and subjected to the very objectionable behaviour of persons who come with third-class tickets and do not spend a farthing (and) wish us to sow for them to reap.' But while the salmon was to be dispatched to a former penal colony, the fish quest appealed to him.

'Poor Ramsbottom was in despair, as his son was anxious to go to Australia with the ova, but as yet they had got none anywhere.' Phelps suggested instead of a net he try with a fly, but 'Ramsbottom shook his head', sure that spawning fish would not rise in such a frost. But Phelps persisted for hours, 'literally cased in ice, my wading trousers as stiff and hard as tin'. Eventually he landed a female of 18 or 20 pounds (eight to nine kilograms), 'a prize for poor Ramsbottom, as she was nearly bursting with spawn'. Then a male, and another smaller female.

'Mr Ramsbottom … was not a little pleased to see the mass of eggs I had for him,' an estimated 4000 ova. Youl was immensely relieved. 'It happened at a most critical time, when the more promising hope of success of getting the living ova of salmon to the Antipodes hung in the balance.'

Youl also won agreement for netting in the Severn and Teme after a magistrate and the Association for the Protection of the Severn Fisheries made it clear policemen would ensure any caught fish was returned to the water. Under the supervision of Frederick Allies, a fishing tackle maker, and with crowds watching, the Severn produced suitable 15- to 18-pounders. William Ramsbottom raced to Worcester to fecundate 40,000 ova on Saturday 16 January, desperately hoping to get them to London in time for Wednesday's departure. On the Scottish borders, large crowds also watched the Johnsons' efforts, aided by the superintendent of the Tweed River Police and local fishermen. Several nettings around the Ettrick and Tweed produced nothing, but estate owner and angler Robert Geddes used rake hooks to grab fish by the back fin. Peter Marshall, keeper of Stormontfield's salmon breeding ponds, was on hand to extract 10,000 ova.

Waiting in London, Youl confessed he was 'driven almost crazy' as the clock ticked. Then, with just 48 hours to go, his prayers were answered: each of his salmon hunters arrived with their bounty. Taking the first available trains from Scotland, Lancashire, Worcestershire and Wales, they each arrived in London on the same morning and rushed to the East India Dock. Youl was much relieved with the estimated tally: 18,000 from the Dovey, 50,000 the Ribble, 40,000 the Severn, and 10,000 the Tweed.

The urgent task now was to pack the delicate ova inside small boxes of common deal, measuring 11.75 by 8.75 by 5.25 inches (approximately

30 by 22 by 13 centimetres) and strong enough to withstand the weight of up to nine feet (2.7 metres) of ice blocks. Kneeling on a cushion, Youl had a pile of soft, moist green moss on one side and a mound of finely crushed ice on the other. In front, a tub of fresh water kept pure with charcoal; behind him, a pile of wooden boxes.

Box by box, Youl first laid a double handful of charcoal to absorb any carbonic acid from the decomposition of any dead ova, a layer of chipped ice and then 'a soft feathery nest' of healthy, freshly cleaned moss. After delicately laying the eggs, he placed a moss blanket over them, and progressively laid more ova and moss until the box was almost full. He poured water in until it streamed from all the box holes, 35 top and bottom, 14 on each side. After a final layer of pulverised ice, the lid was screwed tightly shut.

Onlookers weren't sure if it was magnificence or madness, but they marvelled at the tiny, translucent, amber-coloured globules, about the size of a dried pea with a curious gem-like glow, being sent across the world. One wrote: 'Talk about things worth their weight in gold! Why, each of these is almost worth its weight in diamonds.'

181 boxes were placed on board *Norfolk* in an icehouse about 13 by 9 by 8 feet (4 by 3 by 2.4 metres). It was built of two thicknesses of three-inch deal with an open space of about eight inches, filled with charcoal dust to act as a non-conductor and lined with sheet lead. About 90,000 ova in 164 boxes were placed on a grating on the bottom of the icehouse, about 13,000 ova split between the middle and top.

It was hard work, so cold the two-foot-thick ice blocks froze into a solid mass. In temperatures up to 20°F below freezing (-6°C) with a piercing north-east wind, Youl's fingers were numb and bleeding from handling so much broken ice. When he pleaded it was 'impossible' to meet the scheduled 20 January sailing, *Norfolk*'s owners agreed to wait one day, but it still took 'every exertion' to have the last ova and ice on board by 4pm on 20 January for sailing the next morning.

An exhausted Youl then had a last-minute shock when Frank Buckland and Francis Francis arrived bearing gifts. Youl couldn't believe their offering: the ova of trout, one of salmon's greatest enemies. Aware of Youl's difficulties obtaining salmon ova they had independently sought trout; not the monarch of the rivers, but aristocratic enough, they felt. And being hardier it was more likely to survive and succeed in Tasmania, so if salmon failed he who put trout where it had also never gone before would still achieve fame.

Buckland had approached Sir Henry Keppel, former groom-in-waiting to Queen Victoria and a newly appointed vice-admiral, about 'the urgency' and gained access to a branch of the Itchen running at the bottom of Keppel's garden at Bishopstoke south of Winchester. He found collecting eggs was 'one of the most difficult, and I may say dangerous tasks'.

He listed the requirements: waterproof rubber dress (even though in frost it became 'a suit of inflexible armour'); a spawning tin or bath ('large enough to bathe a good-sized baby in'); a long shallow basket to hold captured fish; lengths of flannel to help hold the fish; dry towels because fish slime made hands slippery ('a very bad thing'); bottles and boxes for spawn; sacks of moss; pincers for removing dead ova; heavy and lighter dragnets; landing nets; and a change of clothes. And a final tip he had learned from the 'esquimaux': a bottle of scented hair oil 'to well anoint the chest, arms and ears'.

He found great difficulty 'having to wade right across the river pulling a very heavy net behind me' but obtained about 1200 ova from common brown trout described as 'regular beauties'.

Francis sourced his ova from 'two of the finest breeds of common trout in England' from chalk streams in the Wye Valley. From the mill pool of Spicers papermakers at Alton, Hampshire, he obtained 800 ova from the Wey, then 700 from a Wye stream favoured by the Prince of Wales at a corn mill near Wycombe, Buckinghamshire. They were 'the finest trout ova I ever saw', from fish of 'wonderful colour and flavour, often of a darker red than salmon ... the very best breed I have ever seen in England'. Buckland initially thought Francis had collected sea trout, of which Francis was a fan – 'one of the gamest fish that swims' – but apologised after Francis insisted they were common trout.

Perhaps they were the best of British, but Youl had been repeatedly warned that salmon's cousin was a predatory cannibal. Ramsbottom Sr told of catching trout with 500 salmon ova in their stomachs, and many said they hoped Australians would never turn trout into the same river as salmon as no fish was more destructive of salmon roe and fry.

Youl took the advice to heart. The tyranny of distance and tropics might be overcome, but not trout. 'I had made up my mind not to send any trout ova because I believed that as the trout grew faster and arrived at maturity sooner and fearing that they would not be able to keep them separate in the breeding ponds, they would gobble up all my little salmon.

'They are the greatest enemies the salmon can have. I can compare them to nothing but wolves in a flock of sheep.'

So when Buckland and Francis presented their trout ova, he was aghast, 'nearly pitching them overboard into the dirty water of the docks'. But he felt it would be 'most ungracious after all the pains these gentlemen had been put to by the severe frost, not to forward them'. He politely thanked Buckland and Francis, but quietly instructed William Ramsbottom to leave the trout in Melbourne as a gift to its Acclimatisation Society.

On the morning of Thursday 21 January 1864, *Norfolk* was towed out in the Thames, loaded with eggs and expectations. For Youl, it marked the possible culmination of a decade's effort to make history. For William Ramsbottom, a voyage which could see fame to match his father. For acclimatisers, 'the ultimate fate of which the attention of the whole scientific world ... was anxiously directed.' For some enthusiastic Australians, the most significant and valuable cargo to Australia since Macarthur's famous merino half a century before, a delivery 'more valuable than gold'.

While *Norfolk* headed toward Gravesend, Ramsbottom Sr and Johnson each took two boxes of eggs set aside to take to their Clitheroe and London homes for possible hatching. Youl took some to the Wenham Lake Ice store, then penned an uncharacteristically florid letter to the *Times*: 'I feel confident your readers will most cordially join with me in wishing the good ship *Norfolk* a safe and speedy voyage, and in hoping that these precious little globules may retain their vitality in their damp mossy bed until they arrive at the sunny clime and golden shores of Australia. So that when placed in their native element they may come forth leaping with delight in the limpid waters of the beautiful river Derwent.'

Supporters said few men would have had the good temper to keep trying but Youl devoted his 'heart and soul ... to such a degree the incessant worry and anxiety have almost driven [him] into a brain fever'. He was worn down by years of hard work, setbacks, and 'ridicules and calumnies'.

If this attempt failed, Youl told friends 'I shall give it up, as luck must be against me, and I must leave the field open to some more fortunate person. I never will again undertake the anxious responsibility of another experiment.'

If it was to be the end, he wanted a final, acerbic word. Writing to the Launceston *Examiner*, he said if his advice on any future attempts was 'thought worthy of asking for' he would gladly offer it, but for now he addressed all those 'who have animadverted so severely upon my endeavors to carry out such a difficult and delicate experiment'. Anyone else trying to achieve salmon success needed to understand that if they should fail, they could expect to be 'judged and condemned publicly by persons who know nothing practically of the subject they write upon, and the difficulties to be met and overcome'.

He had kept records of being derided or dismissed as 'an illiterate scribbler', 'shallow philosopher', 'waster of public money ... on ridiculous experiments', 'untruthful', and with a 'great want of judgment', such that the quest 'never can succeed whilst entrusted to such incompetent hands. In all, deserving of everything that can be said against me'.

He enclosed two letters. Robert Ramsbottom expressed 'unspeakable satisfaction' with the arrangements and his confidence 'success will crown your efforts. Had the management of your experiment been entrusted to myself, I cannot see that in any way I could have improved upon your plan.' William Ramsbottom was 'highly pleased and doubt not that I shall this time land the valuable fish in Australia.'

Youl had done all he could. He just had to wait, for six months, to learn the fate of what he called his 'precious babies'.

13

CROSSING THE LINE

With severe frosts finally easing in favour of 'mild weather', a few degrees above freezing, *Norfolk* Captain Bryan Tonkin hoped telegraphed weather messages would be void of 'gales' and 'storms'. Just the week before, a large steamer *Louisiana* had been forced back after Atlantic gales swept 17 crew and passengers overboard.

Winds were still strong and squally when Tonkin ordered the canvas up as *Norfolk* headed toward Plymouth to collect 39 cabin and 60 steerage passengers trained down from Paddington. The port was famous for historic voyages: Francis Drake's *Pelican* journey in the 1570s, the pilgrims' *Mayflower* to a New World in 1620, Captain James Cook in 1768 on the *Endeavour* to find and possess 'a continent of land of great extent', and 22-year-old Charles Darwin on the *Beagle* in 1831 on a voyage including Van Diemen's Land. On the night of 29 January 1864, *Norfolk* left with scientific, angling and colonial worlds speculating about fresh history.

With the result not to be known for months, Youl could do nothing but wait. As *Norfolk* had run from Plymouth to Australia in 71 days several times, Youl calculated the ship would arrive in Melbourne in April, well within his 90-day life or death limit, with good news perhaps returning to London about May or June.

Supporters were concerned about Youl's 'incessant worry and anxiety', being driven 'almost crazy' with 'brain fever' over his quest and reputation. But the tide was turning. News of the positive ice experiment results saw the *Mercury* now embrace the formerly 'incompetent' as 'an old friend': 'An ordinary man would have broken down under the failure of his two former efforts and have given up the thing as hopeless.' But he was a man who saw difficulties as things to be overcome, and 'the greater they become the greater amount of energy he brings to bear against them … who besides Mr Youl would have had the courage … or the energy to have carried it through?'

Success would 'crown Mr Youl's efforts', it said, but even if the *Norfolk* mission failed 'to what other or better hands could we confide the task? We have gone too far to stop now, until we have made the experiment a complete success.'

The *Mercury*'s former derision was not forgotten. Launceston's *Examiner* hoped a successful shipment would reward Youl as 'a fitting reply to all his traducers', while the *Geelong Advertiser*, edited by refrigeration pioneer James Harrison, said the colonies would owe Youl 'a deep debt of gratitude' as he had 'worked … for nothing, was repaid especially by the Tasmanian papers, with nothing but abuse'.

Waiting for the *Norfolk* outcome, Youl wanted to 'live' the voyage through two boxes of about 1200 ova he had retained. On Wednesday 20 April, about the time he expected *Norfolk* to be in Melbourne, he invited zoologist Dr Francis Day, who was striving to take trout to streams in India, Frank Buckland, Frederick Allies, Thomas Johnson, and the owners of *Field*, to the Wenham Lake Ice Company in London.

With the 'greatest anxiety' the boxes were removed from the icehouse and their lids unscrewed. To Youl's immense relief and surrounding cheers, the eggs were in good condition. Each took eggs home to see if they would hatch. Youl put his into a trough in a garden work shed. On Saturday 23 April, while studying the translucent amber or pink-orange eggs, his eye caught something else: one was on the move, pulling a yolk sac behind it. 'My first little fish was hatched!'

The experience brought alive the words of Buckland in his *Fish Hatching* book: 'I have never yet seen a more beautiful sight than the gradual development of the young salmon. In a globule a future fish begins as no more than a faint line, then the beginning of a head with two tiny black spots, then the beginnings of dorsal and pectoral fins, which seem joined. For about six weeks the new fish relies on its 'forage bag', an umbilical yolk vesicle, before this is shed. Now a fingerling, or parr, no more than two inches long, it begins to take on the salmon's colors, its features gradually emerging as if by magic into the various fins distinctive of the adult creature, and we have a perfect fish before us. Nature, ever wonderful in her works, surpasses herself in the beauty and minuteness of finish of these little fish.'

With Buckland, Allies, and Johnson also reporting 'beautiful little fish' doing well, confidence was rising that ova kept on ice for an extended period could still hatch successfully, at least on land.

At sea, Ramsbottom's experience was entirely different to his previous voyage, but he, too, could know nothing of success or failure until landing. This voyage was nothing like his desperate efforts to save ova on the *Beautiful Star*, with no complex apparatus to manage inside a heaving ship. The ova were tightly packed in boxes inside a sealed icehouse, with strict orders from Youl that it not be opened until Melbourne.

The *Norfolk* manifest was its own colonial tale: reverends, doctors and wealthier passengers in cabins; assisted migrants in steerage. With an inducement of free or assisted passages to rapidly growing colonies needing labour, emigrants were leaving Britain in 'exodus'. Their voyage may have been cheap, but it came with new regulations for all emigrant ships proceeding 'to any port or place in Her Majesty's possessions abroad'. Emigrants were to rise and dress and have their beds rolled up and their deck area swept by 7am. The crew lit oven fires so breakfast could be eaten between 8 and 9. Lunch was at 1pm, supper at 6pm, and everyone had to be in their berths by 10pm.

Strictly prohibited was 'all immoral or indecent acts of conduct, taking improper familiarities with the female passengers, using blasphemous, obscene or indecent language or language tending to be a breach of the peace, swearing, gambling, drunkenness, fighting, disorderly, riotous, quarrelsome, or insubordinate conduct.' No smoking was allowed between decks, and Sundays were to be observed as much as circumstances allowed.

Below decks were colonial 'essentials': thousands of packages of candles, footwear, wine, brandy, chocolate, cheese, pastilles, clothing, hospital supplies, iron, twine and scientific books. And Youl's precious cargo: '1 tank, containing salmon ova'.

Ramsbottom watched passengers doing their best to amuse themselves: sighting whales and flying fish; trying to catch an albatross by entangling its wings in a line of string; staging concerts; playing cricket, vingt-et-un and quoits[1]; singing sacred songs on a Sunday. On 20 February they enjoyed a ceremonial salute to 'crossing-the-line' into the Southern Hemisphere, with sailors dressed as Neptune and Amphitrite, the king and queen of the sea. Then seeing the sun rising and setting across a different sky and spotting the famous Southern Cross.

Replicating the route of the First Fleet and the traditional clipper route, *Norfolk* used the natural winds and currents to run south in the western South Atlantic toward Trinidad, then south-east toward the distinctive volcanic archipelago of Tristan da Cunha, about 1700 miles (2700 kilometres) west of Cape Town. On 20 March, Captain Tonkin might have noted to Ramsbottom they had reached the 38° south latitude, the same as their destination of Melbourne, still 7000 miles (11,000 kilometres) away.

The pair became agitated at even a hint of rough seas. Determined to be the first captain to deliver live ova, Tonkin quickly sought shelter for two days in a south-east gale, the type which threatened shipwrecks and forced three men to the wheel. Whenever a wave thumped the

---

13. 1   'vingt-et-un' is French for 21, Blackjack is a derivative.

side, Ramsbottom immediately thought 'there goes another thousand of them'. He had no way of knowing whether the ova were alive in captivity or dead in an icy crypt. Both were relieved when gales eased and *Norfolk*'s sails were filled by the roaring forties, the trade winds blowing uninterrupted between the Africa and Australian continents, although Tonkin was anxious not to drift too far south, knowing the Admiralty had long recommended captains stay north of 40° to avoid iceberg disaster.

With the world not fully connected for telegrams, a 2 February message from London arrived by ship in mid-March, allowing the *Argus* to declare 'boxes of ova are on their way ... we look out eagerly ... the safe arrival of the precious consignment is looked forward to with much anxiety.'

To the *Argus*, this 'national undertaking' demanded first warship ordered by a British colony, Her Majesty's colonial sail-steam sloop *Victoria*, be on standby to receive and transfer the ova – if it were still alive – on the final leg to Tasmania. The 580-ton (526-tonne) *Victoria* was described on its launch in 1855 as one of the finest craft in the world, featuring Spanish mahogany appointments and the new technology of a feathered screw, allowing propeller blades to fold back when not in use to reduce drag, and a telescopic funnel and steam exhaust pipes which could be retracted when not in use. With three broadside guns, *Victoria* was the Empire's southernmost 'defence', primarily to ensure the passage of Australian gold was not threatened by French or Russian raiders.

But in Tasmania, the government was in 'intense anxiety', fearing Youl had made another fatal error by ignoring instructions that the ship should sail directly to the island. It feared this might prove fatal. 'We are very anxious about it,' the Salmon Commission said, especially 'those of us who are more particularly responsible for the success of the undertaking.' And there was fretting over the 'considerable danger' from *Victoria*'s steam engine tremor, something Youl himself had long resisted.

There was no option but to hope his arrangements, and advice on the transfer from Melbourne, would prove successful. After Edward Wilson and fellow Victorian Acclimatisation Society members persuaded the Victorian Government 'that with a little care' *Victoria* was suitable, Salmon Commissioner and Public Works Director William Falconer rushed to Melbourne in early April to satisfy himself with arrangements.

The small victory in the long battle to have Her Majesty's Navy engage in the salmon mission led the *Mercury* to call for celebrations for 'a great national event'. While *Norfolk* crew and passengers speculated and gambled on their prospective arrival day and time, people on southern shores watched and waited for sightings and any news from telegraph stations connecting the entrance of Port Phillip Bay to Hobson's Bay, Melbourne.

When the anticipated arrival date passed and *Norfolk* appeared on the 'overdue shipping' list, anxiety heightened. But finally, on Thursday 14 April, the vessel was sighted off Cape Otway, the southern tip of Victoria's west coast, the junction of the Southern Ocean and notorious Bass Strait.

Two days later a telegraph was progressively sent 40 miles (64 kilometres) from the pilot base at Queenscliffe to Williamstown, Hobsons Bay, and finally to Melbourne: '*Norfolk*, ship, from London'. On a fine autumn morning, with the tide high, the vessel anchored in Hobsons Bay, 85 days after leaving the East India Docks, 77 since Plymouth.

Wilson and other Acclimatisation Society members rowed out, anxious to learn whether the voyage was historic or another failure. Everyone held their breath as Ramsbottom unlocked the icehouse. The first positive sign was that only about a third of the ice had melted. It was a fillip for the nascent refrigeration industry – James Harrison had just patented a pioneering machine and would soon use the same *Norfolk* for an experimental refrigerated beef shipment from Australia to the United Kingdom – but all eyes were on what lay beneath the ice.

With 'fear and trembling' Ramsbottom unscrewed a box lid. It was a 'profound sensation' when at least 80% of the ova appeared to be in perfect condition. Ramsbottom discerned the eyed stage and proclaimed that in 21 days the first baby salmon would emerge from its embryo bag and 'come to life'. He decided there was no time to waste, no need to inspect more boxes. *Norfolk* was immediately pulled into the railway station pier at Sandridge while Commander William Norman steamed HMS *Victoria* across the bay from Williamstown.

The salmon ova arrival was enough 'to send the disciples of Izaak Walton into ecstasies', according to the *Farmer's Journal* and *Gardener's Chronicle*. And it was one of two arrivals of quintessential English culture.

HMCS *Victoria*, first warship ordered by a British colony. Courtesy Australian War Memorial, AN 300060

A second All-England Eleven cricket team was delighting crowds on a tour of Victoria, New South Wales and New Zealand. The players, led by George Parr, known in his day as the greatest batsman in the world, were undefeated against sides featuring Tom Wills, credited with being Australia's first cricketer of note, and founder of Australian football.

But it was the prospect of other 'parr' which led the Melbourne *Herald* to declare the excitement at Windsor on hearing the news of a son born to the Prince of Wales was nothing compared to the startling intelligence that salmon had been delivered. *Bell's Life* and *Sporting Chronicle* salivated at the prospect of 'salmon steaks, salmon with capers and salmon du bleu'. The Acclimatisation Society forecast new occupations in making hooks, flies, lines, rods and boats, with the colonies emulating Scandinavians, Venetians, Genoese and English with valuable fishing trade and commerce.

A *Sydney Morning Herald* contributor said, 'all the world are doing homage to the 'little strangers'.' The Melbourne *Punch* was jubilant: 'Hooray, hooray! we've hooked our fish/By Jupiter of Ammon/As sure as ever eggs are eggs/We'll hatch colonial salmon ... Our hungry praise we utter/And salmon see – colonial bred/Swim in colonial butter.'

People were catching, trading and eating salmon before the eggs had even arrived in their nursery, let alone hatch, mature, migrate, breed, reach adulthood and be caught. The salmon distribution, and the fate of the 'accidental' trout, was in dispute. Victoria's Chief Secretary and Premier, the Scottish-born James McCulloch, and Wilson's Acclimatisation Society, felt it prudent to leave some salmon ova in Melbourne to guard against 'all the ova being killed in their passage across to Hobart Town'. Especially, they said, when Youl himself had written 'again and again' about his 'dread' of a steamer.

The Society also cited Victoria's support for the salmon quest, and that without the provision of *Victoria* 'it is probable that the entire success of the experiment would have been endangered'. Anyone who thought leaving some ova in Victoria was not reasonable was 'ungracious and childish'.

But Ramsbottom argued Tasmania had long been nominated as 'the first home of the salmon' and sparing even a few hundred ova was not prudent. Breeding salmon in Britain was difficult enough with hundreds of thousands of salmon, so it was 'unwise to throw away a chance of success by withdrawing even 400 fish from our small stock'.

It was finally agreed Victoria would retain eleven boxes in a Harrison icehouse in North Melbourne until Ramsbottom could recommend where any hatchlings might be transferred, perhaps to the Snowy River, Gippsland or upper tributaries of the Yarra.

As for the 10 boxes of trout ova, Youl had given Ramsbottom clear orders to leave them with the Victorians and quietly telegrammed the Victorian Acclimatisation Society about their 'gift'. But Morton Allport had pressed on Ramsbottom the merits of delivering trout, as 'the salmon was at best doubtful and [trout success] would be absolutely certain if we could get living eggs.' He had previously told Ramsbottom that if Youl eschewed the pursuit of trout for official shipment, he was authorised to spend £50 on his private account for 'purchasing and packing 1000 trout ova for myself'.

Ramsbottom, confident he could 'nurse it into a success' in Tasmania, won the day, much to Allport's eternal satisfaction. With the matter resolved, it was time to quickly transfer 170 boxes of salmon ova, and the trout boxes, to the lower hold of *Victoria*, along with 12 tons of remaining ice.

Youl had arranged for 11 large wooden crates to each accommodate about 15 boxes of ova. The boxes were surrounded by ice, sawdust bags and blankets to minimise melting, with pads wedged between to mitigate movement. The *Herald* hoped 'the dear little creatures' would not be killed by over-kindness.

The hard work was unfinished when it was too dark to continue and resumed at first light on Sunday 17 April. Finally, at 2pm, the last of the ova was safely aboard. A Hobart verse publication, *Salmoniana,* said that on *Victoria* 'that night ... the great men talked science and salmon in fine, how they smoked, laughed and drank all the captain's best wine'.

But anglers well knew there was many a slip between a sighting, a bite and final landing. The delicate eggs still had to be safely transported across one of the world's most treacherous stretches of water, unloaded at Hobart onto a third vessel to head upstream in the Derwent, then be trekked overland to ponds on a remote tributary.

'We confess to a feeling of exultation', said the Launceston *Examiner,* 'but all has not, however, been accomplished yet. There are difficulties and dangers yet to be encountered before we have breeding salmon in our rivers, and until then the risk will not cease.'

14

WELCOME LITTLE STRANGERS

Hobart buzzed as word quickly spread of the imminent arrival of Her Majesty's 'sloop of war' making an inaugural visit to deliver the miracle of long-awaited salmon and unexpected trout.

Ramsbottom and Captain Norman were relieved the 500 miles (800 kilometres) of Bass Strait presented none of its frequent treachery, but head winds and fog made for cautious progress for two days before *Victoria* was spotted heading up the Derwent from Storm Bay.

Moving to anchor off Battery Point in the mid-afternoon of 22 April 1864, the ship was surrounded by numerous small boats and dinghies, decorated and filled with eager and curious occupants, and trading barques dipping their ensigns. Despite newspaper calls for loud rejoicings of a 'national event', a celebratory firing of Hobart's battery gun was forbidden because of fears it would send a fatal tremor through the ova.

As a crowd massed on shore, anxious members of the government, Salmon Commission, Royal Society and Acclimatisation Society rowed out to *Victoria*. Ramsbottom quickly showed his own 'little scheme' of keeping 50 ova in a saucer of moss and ice. Only a handful had died on the voyage, a 'wonder' to the official party. Commissioners reportedly gazed at the ova 'with tears in their eyes', excitedly exclaiming 'By jove! ... released from their bonds, the salmon shall swim in the ponds.'

The unloading of eight 'native bears' (koala) from *Victoria* for the Tasmanian Acclimatisation Society was another attraction, but the arrival of salmon caused nothing less than a 'sensation'. Supervised by Ramsbottom, Norman's men quickly began transferring the large containers of ova boxes, and remaining 10 tons of ice, onto a waiting barge. After six hours of hard work the task was completed, and at 9pm the barge was taken in tow by a small steamer, *Emu*, which connected Hobart with the village of New Norfolk, the landing point closest to the salmon ponds prepared two years before.

Under bright moonlight, *Emu* proceeded 22 miles (35 kilometres) upstream. At Bridgewater, where the Derwent's upstream fresh water blended with salty estuary water, rousing cheers could be heard from spectators at the Black Snake Inn. The timber bridge span was winched up to allow *Emu* to pass, watched by curious native black swans.

Then panic set in. A leak had been discovered. Everyone feared the worst, the real 'possibility of ... disastrous results' if salt water reached the ova, or worse a sinking or major delay, but the barge operator assured everyone the leak was small and his baling would save the day.

At 1am *Emu* reached New Norfolk, the island's third oldest settlement, named by early settlers who had been compulsorily transferred from Norfolk Island to make way for a notorious convict settlement. Despite the late hour, excited locals turned out in force. Many wanted to celebrate with cannon fire but were sternly told a welcome volley could 'bust all the eggs'. The 'invaluable cargo' was securely moored and placed under guard for the night.

The next morning, Ramsbottom could see the Englishness of New Norfolk, with life built around a village green, some of Australia's earliest licensed inns and an 1824 Anglican church with impressive stained glass. But also, reminders this remote 'little England' 'had penal history, with soldier barracks, convict cells and a lunatic asylum.

Ramsbottom had no time to admire the willow-banked Derwent, flanked by pioneering hop grounds and orchards, as Salmon Commission chairman Dr Officer and other influential settlers had arranged for 50 servants and farm labourers, 10 teams of horses, and numerous carts to be at his disposal at 7am.

The first task was to tow the barge from the steamboat jetty to a point better allowing the transfer of the 11 large containers and eight tons of remaining ice onto land. Bushy Park hops-orchard pioneer Ebenezer Shoobridge had four labourers rowing a small boat to pull the barge upstream. Watched by crowds on the banks, the hard work demanded another boat crew join in until, after two hours, the barge was moored about three miles upstream alongside a jetty at Ark Inn, the termination of the navigable portion of the Derwent marked by exposed bedrock known as The Falls.

The salmon ova, or 'little strangers' as newspapers described, were now in the Southern Hemisphere, on Tasmanian soil, in the same week Youl had opened his ova box in London. The remarkable moment was captured in *Salmoniana* verse: 'A motley crowd, more varied than that/ Was ne'er landed by Noah on Mount Ararat;/ Horses, carts, ponies and dogs and cats/ Well-dressed women and squalling brats/ Mischievous urchins, and how shall we style 'em?/ Lunatics from the New Norfolk asylum/ Labourers gathered from farm and store/ To see the live Salmon come safely ashore.'

Laborers transferred the ice into seven horse-drawn carts for a four-mile (6.4-kilometre) trek to the breeding ponds. The task of transporting the large containers with ova boxes inside was handled by feeding 12-foot (3.7-metre) bamboo sticks through rope handles to allow four men to carry each container on their shoulders. Amidst the power of a steam-age industrial revolution, the final delivery was by horse and cart and the same carrying style employed by Chinese carp farmers in 3500 BCE. A reporter wrote: 'A spectator could not help being struck with the idea that he saw a great scientific achievement, culminating up to that point in one of the rudest forms of mere physical effort.'

In relays, the men moved as briskly as they could without rocking their load on rough footing for two hours. With the first containers deposited, they returned to the Ark Inn jetty to fetch more.

Ramsbottom had raced ahead to check the ponds. Under urging from Youl, the first decision of the Salmon Commissioners after their appointment was to favour the east bank of the Plenty, about two miles (3.2 kilometres) from its junction with the Derwent, as the preferred ponds site. It offered ample volume, low temperature, a gravelly bed, and proximity to the navigable portion of the Derwent. It was also inaccessible to poachers' nets because of rocks, foliage, and steep banks. There were no known predatory fish above the Derwent's tidal influence, and, unlike many British rivers, no locks, weirs or industry pollution.

The ponds were on the Redlands estate of hops pioneer and Salmon Commissioner Robert Read. His impressive homestead, built by convicts, was surrounded by plantings of English oak, chestnut, poplar, beech, maple and elm. After joining the East India Company as an 11-year-old midshipman, Read capitalised on Tasmanian shipping trade and an 800-acre (320-hectare) land grant. Locals speculated 'The Squire of Redlands' was an unrecognised outcast son of King George IV, whom the *Times* had observed would always prefer 'a girl and a bottle to politics and a sermon'.

About half a mile (800 metres) past the homestead, a hawthorn archway marked the entrance to the hatching site. It was on three acres of land which Read agreed to rent for £15 a year, along with providing a common way along the banks of the Plenty and access to the convict-built channel irrigating his fields. He also housed Ramsbottom to save him from freezing in a tent.

God's hand had seemingly created the idyllic location. The *Mercury* enthused the Plenty was 'the very beau ideal of a salmon stream', and British readers were told that while many mainland rivers often had too much or too little water for the monarch of the stream, 'the Prince of Australian rivers for sustained beauty and grandeur is the Derwent, from its cradle to the grave'.

The cradle was in the very heart of an island born of millions of years of up to five glaciations, with ice up to 1640 feet (500 metres) thick breaking a continuous land mass to create Tasmania's separation from continental Australia. In 1793 Lieutenant John Hayes, on an exploring expedition for the British East India Company, christened the River Derwent after the Derwentwater and Derwent River in the Lakes District of his native Cumberland. Two months earlier French explorer Admiral Bruni D'Entrecasteaux, searching for the lost La Perouse expedition, had named it Riviere du Nord as the river flowed from the north.

Neither knew the river began as a trickle in the Central Highlands at Lake St Clair, Australia's deepest lake. George Frankland, who led a survey expedition to find the Derwent source in 1835 and hosted Charles Darwin's 27th birthday celebration the following year, thought the 'enchanting beauty equal to the little cantons of Switzerland'. Others felt 'it is easier to imagine oneself in Scotland or Norway than at the Antipodes'. But while famous Scottish lakes were in the valleys, exiled Irish political leader John Mitchel saw those of Tasmania as 'hung up among the clouds', the pristine waters 'high above all the odious (convict) stations and townships ... the whole world of convictism and scoundreldom.'

The splendour drew Frankland to christen the landscape after Greek mythological figures and gods, such as Mounts Olympus, Ossa and Acropolis. From the peaks the Derwent descended more than 2300 feet (700 metres) over 148 miles (238 kilometres) past New Norfolk, about as wide as the Thames at Windsor, then expand past Hobart to nearly two miles (3.2 kilometres) wide.

The ponds were near the Derwent's midpoint, where the Plenty River flowed in from the land of giant ash, the world's tallest flowering plant, at Mount Styx, named for a mythological river separating the living from the dead.

The hatchery, which cost £727 and an additional £60 for fencing, was based on what Morton Allport's brother Curzon had sketched on a visit to Stormontfield, supplemented by a *Times* description sent by Youl. Ancient Huon pine, found only in the rainforests of South-West

Tasmania, was sourced for construction, the timber highly valued as its oil content rendered it waterproof and impervious to insects. A large wooden race carried water from the Plenty to a grassy area above the estimated flood mark, with sluices to regulate the flow. A smaller race ran off to feed water through smaller sluices into a circular pond about 35 feet in diameter and five feet deep (10.6 by 1.5 metres).

Pond water then ran through a series of closed and open wooden troughs and more sluices into 15 gravel hatching beds. The ova beds of gravel, set in five rows, were about five by two feet deep (1.5 by 0.6 metres), subdivided into smaller compartments. With a slight fall, water entered the first box in each row, passed to its lower end and onto the upper end of the next, and progressively down each box.

Finally, the water passed to a large pond about 130 feet long and 40 feet wide (39.6 by 12 metres), ranging in depth from two to nine feet (0.6 by 2.7 metres). Access to Read's pioneering irrigation ensured an independent back-up to the race system and a constant supply of Plenty water. All water entrances and exits were covered with perforated zinc gratings to guard against predatory insects or animals.

If hatching was successful, young fish would swim from the ponds into the Plenty and then the Derwent and hopefully make their way into the salty expanse of Storm Bay and the Tasman Sea.

But Ramsbottom knew that day was a way off. And on arrival day he was not happy. The gravelled hatching trays built two years before were now full of sediment and impurities, making for a salmon grave, not a nursery. To the disappointment of the crowd of officials, labourers and spectators, Ramsbottom insisted every piece of gravel be individually cleaned. So everyone set to washing for several hours, until, in the late afternoon, he was satisfied he could prepare a nursery of redds, or nests, with clean gravel.

The water temperature was another source of anxiety. The ova had been taken in a freezing British winter and transported in an icehouse. Now they faced unusually warm autumn temperatures, with Plenty water at 54°F (12°C). Ramsbottom estimated he now had just eight of the original 30 tons (27.2 tonnes) of Norfolk ice, with no ice works in

Hobart for replenishment. He hoped he had enough blocks to place in the wooden compartment running over the breeding trays to keep the temperature between 39°F and 49°F (3–9°C).

Then, finally, came the moment everyone was waiting for for: time to lift the lid on salmon. Under a tent to mitigate the sun, Ramsbottom and Allport carefully unscrewed the first of the boxes in which ova had been imprisoned.

Ramsbottom admitted to 'intense anxiety'. Youl had always represented 100 days as the limit after which 'it would be highly dangerous to delay their immersion in their native element'. It was now 96 days since most of the ova had been taken from the parent fish.

Lifting the first layer of moss, Ramsbottom was in despair, much of the ova dead or dying. 'The first couple of boxes ... were calculated to give rise to the most gloomy forebodings,' a *Mercury* reporter noted. But as more boxes were opened the gloom began to lift. With ova of 'decidedly more encouraging character', the reporter rushed back to Hobart to write his report, which would be forwarded to the *Times*. As not all the boxes had been examined, he wrote, it would be premature to judge how many had survived, but 'it is ... certain that there is a quantity of healthful spawn now actually deposited in the hatching boxes to stock, provided that an equal proportion of success to that experienced in the old country attends the process of incubation, not alone the Derwent but many other Tasmanian rivers with the family of the noblest fish in existence.'

Work continued for the rest of the day, and by candlelight through the night. The unpacking was even more laborious than Youl's packing. With each layer of moss, Ramsbottom used fine pincer sticks and tweezers to remove any dead or unhealthy ova, striving not to injure survivors. He could see the health of the ova clearly depended largely on the moss: where moss retained its natural green sponginess, most of the attached ova were healthy, but few survived where moss was brown and compressed. Ramsbottom then gently lowered a moss layer upside down into one of the breeding trays, allowing ova to drift away into cool water.

The work was completed by nightfall the next day, Friday 22 April, 1864. There was now salmon ova in the Southern Hemisphere for the first time, but Ramsbottom knew this was just the end of a beginning. Final success depended on ensuring the ova hatched and survived.

Ramsbottom paid close attention to the boxes of trout ova, conscious that while Allport declared trout most welcome in Tasmania, Youl saw it as salmon's worst enemy. Most were dead but he retrieved a small number and kept them well separated from the salmon.

For days Ramsbottom and an assistant, John Stannard, maintained constant surveillance for any dead salmon ova – initially about 100 a day – and any lingering moss strands, or minute elements of charcoal or straw. All had to be carefully removed.

When Salmon Commission chairman Officer pressed Ramsbottom on the number of living salmon ova, he hesitated before estimating 30,000, less than a quarter of that which left London, and perhaps 150 trout.

The numbers were somewhat disappointing, but Officer raced to pen a letter to Youl. He could not give any accurate figure but was satisfied there were 'many thousands ... amply sufficient, if they should all continue to thrive and should become living fish, to ensure the complete success of our experiment.

'No human first born was ever ushered into the world with more rejoicings than these will be.'

The colony had been built on the backs of convicts separated from their families, transported across the seas and held captive on foreign land; now it welcomed a new future with salmon and trout ova, separated from parents, transported in captivity in sealed boxes, and held in alien water.

Youl had long waited for this alchemy of practical and new science. It promised another special bond between Mother Country and her Antipodean children, a new angling and dining pleasure, and perhaps a new industry for Tasmania. In all, a resounding riposte to all who had echoed the words of writer-angler Henry Kingsley that 'we consider the chance of introducing salmon into Tasmania as a hopeless business'.

Only now did former detractors appreciate 'how slender a thread the success ... hung'. The *Mercury* saluted 'the toil and anxiety' Youl had gone through and was relieved he had not 'abandoned the task as an utterly hopeless one'. It trumpeted 'one of the greatest triumphs of modern science'. After being infamous for its convict origins, salmon now gave the island a chance to 'become famous for that, if nothing else'.

Friends and supporters wasted no time patting Youl's back. Lachlan Mackinnon told him 'no other man whom I know would have stuck to the matter through evil ... as you have done. When (John) Macarthur and his merinos will be so mixed up together as that you can't separate the grain from the chaff, Youl and his salmon will go hand in hand down to the last of posterity in Australia'. William Archer congratulated him 'on the success of your grand salmon experiment ... we are all eating salmon in imagination.'

Edward Wilson declared: 'Your statue is secure. It will be erected over the future fish-market in Hobart Town, and you are to be in the attitude of Medicean Venus, with a salmon instead of a dolphin. At any rate you may congratulate yourself upon a brilliant triumph, and the fact that you have not lived in vain.'

The usually sneering *Punch* joined the acclamation: 'Hurrah, Hurrah! we've hooked our fish/ by Jupiter of Ammon/ As sure as ever eggs are eggs/ we'll catch our colonial salmon... the credit rests upon/ the fellows who sent out the eggs/ and those who egged them on.'

In Britain, the *Daily Telegraph* said it was 'gratifying to announce that at length that important success so often attempted ... the introduction of the salmon ova into Australia has been accomplished'. Scotsman editor-writer Alexander Russel noted the contrast between 'Mother' and child when it came to salmon. 'Great nations of the future, like Australia and New Zealand, are labouring to obtain what we have been conclusively losing, or even wantonly destroying.'

Bell's Life in Victoria declared the arrival of salmon – 'more valuable than gold' – meant the 'crisis is over'. All 'anglers and bon vivants, whose hearts have palpitated with feverish emotion' could now banish all anxiety. These salmon 'babies' would soon offer them sport and luxury.

But the king of fish could not be saluted yet. While Youl's quest was, the *Argus* said, 'one of the boldest efforts ever made in the service of acclimatisation', it urged people not to 'savour counting our chickens before they are "hatched".'

Much depended on Ramsbottom's midwifery, and he would well and truly earn his £150 salary if he was to meet the hopes of all those watching in Tasmania, other colonies, and around the world.

15

HATCHING HOPE

Youl must have treated himself to a heady draught of satisfaction. He had seemingly delivered a miracle of fishes, and the news was eagerly consumed throughout Australia, New Zealand, Britain, France, North America, India and South Africa.

And it led to a pilgrimage to the Plenty for those wanting personal witness. A red-coated coachman and his bugle called passengers to the Perseverance coach leaving Hobart's Albion Inn at 8am daily. Four greys took passengers on the southern side of the Derwent where wild bush was interspersed with English-looking farms, cottages and gardens before a break at the Bridgewater Inn for a change of horses and a glass of ale. Visitors held their breath as the trek took them close to banks descending steeply to weeping willows.

In the heart of hop country, lunch was taken at the Star and Garter Hotel in New Norfolk, before visitors hired horses or gigs for the final

The salmon ponds. Illustration by William Piguenit, lithograph published by ML Henn 1867, Hobart. Courtesy National Library Australia, CD-20873856.

seven miles to the Salmon Ponds. At the end of a trying seven-hour journey, visitors could see what one described as 'just the mysterious, shady, inapproachable sort of stream' which drew disciples of Izaak Walton to venture with wicker baskets, fishing rods, lures and sandwiches 'in quest of salmon'.

Regardless of the weather, visitors found Ramsbottom working day and night. He had to constantly remove any dead or ailing ova, pieces of moss or any other impurity in the hatching boxes and assiduously monitor the trays and ponds. He also took 'great pains to secure his prize from enemies ... acquiring a taste for such delicate morsels'.

Whether the ova would survive, and hatch, remained in anxious suspense, no one knowing whether the Southern Hemisphere was too alien, and 42° south a mirror of the Northern Mediterranean latitudes where most said salmon were never caught.

The suspense ended and a new chapter the story of salmon began on Wednesday 4 May 1864. Ramsbottom was carefully scanning salmon and trout ova in their separate ponds when he noticed an empty trout egg case. Nearby he saw something wriggling. The first egg had burst! To his immense excitement he could make out a trout fry – the first born in the Southern Hemisphere.

The next day, Ramsbottom was even more exhilarated: the first of the venerated salmon had hatched. The news was rushed to Hobart

Map showing Salmon Ponds site at Plenty, north-west of Hobart

where the *Mercury* and *Advertiser* excitedly headlined 'The First Salmon' and 'First Live Salmon'. The island at the bottom of the world was on top of the world. Rule Britannia, Rule Tasmania! For years to come, the *Mercury* said, May 1864 would be 'forever memorable as the month of the first demonstrated success'. The time when 'Man, once again, triumphed over Nature'.

Then Melbourne was in rapture on 7 May at the North Melbourne ice factory with the emergence of the first salmon on the continent. The *Argus* enthused: 'No more important birth has taken place in the colony ... have we ... had any so important birth?' No 'stranger' was as welcome as the 'unsightly little animal now wriggling about in ludicrous activity and bustle within ten minutes' walk of this office!' No birth had yet been recorded that was 'calculated to exert so wide and beneficial an influence on our social wellbeing as this'.

So many rushed to see 'the most important nativity that has occurred in the colony' that visits had to be restricted, the excitement enough for some to call for 7 May to be instituted an annual public holiday.

Any new supply of food was important, the *Argus* said, but one could not over-estimate the significance of the colonies having the king of fishes, which 'the lowest as well as the highest can enjoy the sport of catching ... and the pleasure of eating'.

Bell's Life in Victoria applauded Youl's efforts as 'another proof of the truth of the old maxim, that perseverance overcomes all difficulties'.

All those 'whose hearts have palpitated with feverish emotion ... may now with a sigh of relief banish all care on the matter' andanticipate days of sport and luxury.

Melbourne's *Punch* lauded 'Salmon see, colonial bred, swim in colonial butter ... let us sing to Wigram's praise, with those who scorning mammon/ brave workers to successful end, helped half the world to salmon.'

Ramsbottom recorded the last salmon, about the size of a fingernail, escaping its shell on 8 June. To his 'surprise and gratification' it became difficult to count the tally. He got to 1000 before it became impossible as the little creatures' instinct saw them wriggle away out of sight under gravel to escape detection. But he hoped the number 'may exceed rather than fall short of expectation'.

He was less optimistic when he visited Melbourne. Fewer ova than expected had hatched in the ice factory and his chosen holding site was an arduous 45 miles (72 kilometres) away. Nevertheless, at 3am, he took a 12-hour coach ride to Badger's Creek, about a mile from its junction with the Yarra, with about 160 hatchlings in a 'travelling tank' swinging from a universal joint, constantly cupping water in and out to keep it 'aerated'. With the help of local Indigenous people, the hatchlings were placed in a perforated 21- by 6-feet 'tank' (6.4 by 1.8 metres) on the bed of a stream to await any growth. Ramsbottom strongly urged Victorians to appreciate 'it's only an experiment'.

A month later, Youl's English summer began with a telegram from Edward Wilson advising of signs of life in the Salmon Ponds at New Norfolk. Youl immediately penned a letter to the *Times*, saying this was 'most gratifying intelligence' to everyone who had helped him secure ova when 'I had almost despaired'.

Ramsbottom also wrote: 'I am pleased to say that I have about 3000 young salmon, all in first-rate order, and as lively as possible ... one of the nicest sights ever seen in this country.'

He estimated there were 50 trout fry in a separate pond, but admitted at least seven had somehow found their way into the salmon pond. They were doing even better than the salmon, getting plump on morning and evening meals of boiled liver. Notwithstanding Youl's 'cannibal' fears,

Ramsbottom considered the trout alone to be worth all the expense, although some suggested they be moved, perhaps to North West River Bay, to give the salmon 'a better chance'.

Robert Ramsbottom, who had once told Youl he might as well fetch Australia to England as to transport salmon ova across the world, was delighted his son was part of such historic news. 'No one can tell how happy I feel.' He congratulated Youl and sided with him on his breakthrough: 'I shall never attribute the ice question to any soul but yourself, and I should not like you to suppose anyone to rob you at that honour.'

Youl was mightily pleased to tell Britain, 'it is an established fact that salmon and trout can be sent to the Antipodes and hatched there'. But disappointed that 120,000 ova loaded on *Norfolk* had been reduced to 30,000 embryo and then only 3000 fry. Especially when for each 100 ova he had kept on ice in London for 100 days, some 80 had hatched. But Ramsbottom and the Salmon Commissioners felt 3000 fry quite sufficient to stock the Derwent and solve the grand problem of whether Australian waters could be home to British salmon.

Youl wondered whether the toll was due the tedious difficulty of removing ova from moss, its sudden exposure to light after 93 days in darkness, the Plenty water temperature being prejudicial, or the constant checking and removal of dead ova causing more harm than care. He didn't think the ponds were an issue, based on four 'stereographs' of the site, sent by Allport to *Field*, which described the Plenty as 'a very charming looking little river ... strongly reminding us of some well-wooded Welsh salmon and trout rivers.'

The London *Times* correspondent reported 'our great salmon-hatching experiment has ... been hitherto remarkably successful', with thousands of tiny salmon, called parr, alive and active. 'Whether they will ever get above par – excuse the involuntary pun – time alone can show. Naturalists here are divided on the subject.'

Field was delighted the 'British character for perseverance and determination has not suffered by a temporary transplantation to the New World.' In the teeth of failure after failure, with no material reward, it took the pluck, energy and persistence of Youl to achieve success, and if any distinction could be bestowed on him no countryman was more deserving.

While Youl had initially been angered when he learned trout ova had been landed with salmon, enough people in England persuaded him this was itself a major achievement, and the trout would succeed even if the salmon did not. He generously told *Field*: 'If ever the disciples of Izaak Walton should have the pleasure of catching a trout in Tasmania they will be entirely indebted to (Buckland and Francis) as I had made up my mind not to send trout ova with the salmon because I believed that as the trout grow faster they would gobble up all my little salmon before they got to salt water.' He made no mention of Morton Allport persuading Ramsbottom to ignore Youl's request to leave the trout in Melbourne. Allport saw himself as the trout hero, telling friend John Gould in London he 'made' Ramsbottom deliver trout 'although this was strongly opposed by those conducting the experiment in England'.

Before any salmon hatchlings survived to migrate to sea, where many in Britain feared they would fall victim to predatory fish, sharks or whales, Ramsbottom had to become ova protective. Allport warned of the platypus, 'the beast with a bill', while Henry Button, co-owner of the *Examiner*, warned of the indigenous beaver rat, sometimes known as Australia's otter, 'a very destructive little creature ... an inordinate fish-eater'. Ramsbottom and Stannard used guns and four terriers on day and night protective patrols.

War was also declared on black-faced cormorants, or shags. The birds, good divers able to hold their breath for up to 60 seconds, were labelled 'brutes' by Ramsbottom's father because they could swallow 50 or 60 fry in a single meal. So, his son shot as many as 200 on a single day, and just as island authorities once put bodies of hanged criminals on a gibbet as a public warning about the consequences of straying from the 'right path', he set cormorant bodies on every available post and fence.

As winter set in, Ramsbottom's efforts were rewarded. The grateful Salmon Commission appointed him 'superintendent of the ponds', with an increased £300 salary (£32,000 today, A$67,000). And the government agreed to spend £100 to build 'a comfortable weather-boarded cottage' for his family when they arrived from England.

With the fate of the first hatchlings unknown, Ramsbottom was determined they should not be the first *and* last. Until the colony had its own 'breeders', he said, it would be 'one of the most foolish things possible' if the government did not seek more salmon ova. It would be worse than a farmer growing one year's crop, and then neglecting to sow his seed for the next three years in the hope one year's produce would be sufficient for all his wants.

Button promoted the call in his *Examiner*. Even if the first ova reached maturity 'any unforeseen casualty could see them all carried off'. So, it was 'very desirable – nay it is indispensable' that further shipments be obtained for the next two seasons. 'To suspend operations for that period would be folly.'

The Salmon Commission persuaded the government to spend another £800 (£86,000 today, A$180,000) to secure another batch 'so that no means will be left untried to bring to a successful conclusion this great experiment'. It was, they argued, now more commercial transaction than experiment, and with the 'additional experience' gained by Youl, 'themselves', and Ramsbottom, more favourable results could be expected.

The Commission sent a letter to Youl soliciting 'his valuable assistance', saying his 'co-operation they regarded as almost indispensable to success', and told the government it did not doubt 'his ready compliance'. But Youl was not so compliant. He declined to 'engage again in a work which had already cost him much personal labour and anxiety' on a voluntary, unrewarded basis for those who had not been his friend.

The *Australian and New Zealand Gazette* said, in words that could have come from Youl himself, no one could have had any idea how much time, expense and difficulty would be involved in the salmon quest. Success was 'solely due' to his triumph over 'numberless difficulties that have worn out the patience and crushed the hopes of every other person engaged in it'.

Youl had received widespread support and praise in England, Scotland, Ireland, Wales, Britain and France, with a broad view that there could hardly be a more deserving countryman on which to bestow distinction. But distinction had not been bestowed in Tasmania, where he received what he catalogued as 'abuse', 'ingratitude' and 'lack of credit' from those ready to congratulate themselves or praise others. Friend William Archer told the *Examiner* the Salmon Commission owed Youl more than their expression of appreciation for his 'persevering and judicious exertions'. To him 'entirely belongs the credit' of the first experiment with *Beautiful Star*, further experiments in the vaults of the Wenham Lake Ice Company, and all the arrangements for the *Norfolk* voyage. Yet the Commissioners made no mention of Youl's preparation of the successful moss-box shipment, when 'if it had not been for Mr Youl this experiment would not have been made at all, and we might have expended much money, and a great deal of time, before we succeeded in introducing salmon into the rivers of Tasmania.'

Youl's rejection of the request to steward another shipment presented the Commission with what it called a 'difficulty'. It asked the Acclimatisation Society of Victoria to try to persuade Youl, but he again declined, saying he had superintended three shipments and 'shrinks from the labour and responsibility of attempting a fourth'.

The Commissioners then turned to Edward Wilson, back in England for more eye surgery. But he and Youl had come a long way together, and Youl tersely told the Salmon Commission that doctors had ordered Wilson to rest after a painful operation and he was not, and could not be, their agent. But Youl said he would accommodate the Commissioners' request to help his 'old friend'. The Commission said it was 'relieved from all embarrassment by the spontaneous offer of Mr Youl again to undertake, on behalf of his friend, the whole management of another shipment of ova.'

While having previously declared *Norfolk* his final mission, Youl was now determined to keep pushing salmon, and his legacy, over the line.

The *Times* welcomed that for 'the first time in man's history ... a regular interchange has been attempted between the living products of different countries to the world'. Like many in Britain, it remained part-incredulous and part-admiring of the extent of efforts to populate Australia with sheep, cattle, camel, goat, deer, hare, rabbits, pheasant, partridge, duck, geese, pigeon, dove, thrush, blackbird, skylark, starling, sparrow, myna, robin and canary. Despite 'zealous efforts', however, the *Times* said they had failed to show they had obtained and sustained salmon.

Others were more positive, seeing the Tasmanian efforts as showing a 'child' could teach its parents. The provision of £800 for another shipment 'shamed' Britain's propagation efforts, as on a population basis it equated to £250,000 (£26 million today, A$53 million). Leading Scottish editor Alexander Russel lamented that settlers in a faraway colony were more devoted to the monarch of the stream than those in the salmon's main kingdoms. *Harvest of the Sea* author James Bertram said they were labouring to obtain what a 'delinquent' Britain was 'carelessly losing or even wantonly destroying', neither protecting nor growing fish. Frank Buckland complained his countrymen seemed to be forever in 'fatal pursuit' of salmon through mills, weirs, steamboats, poachers 'no better than barbarians', and pollution.

Bertram decried those who could not see 'salmon culture would in time become as good a way of making money as cattle feeding or sheep rearing'. Youl, with support from others, had shown what was possible. 'Those very wise people who never do anything but are largely endowed with the gift of prophecy ... proclaimed it could not be done; that it was impossible to take the salmon out to Australia.'

Having a Northern Hemisphere fish in the Southern Hemisphere for the first time had already turned the world of salmon and trout upside down, and, a British Association meeting in Bath was told, if local salmon stocks continued to be annihilated, there would need to be 'trials to reimport from Australia the ova of the offspring of our English salmon.' This was way beyond what anyone had ever anticipated.

16

FOSSILS & PEARLS

Youl received a Christmas gift from Governor Thomas Browne: two preserved young fish hatched from the *Norfolk* ova, which *Field* displayed in its office window with an endorsement: 'The young salmon are very fine species of their race.'

But small hatchery fish in a London window was not proof of salmon in Hobart waters, and elite men of science remained resistant to the efforts of 'practical' men. John Gould believed that if the rivers and seas of Tasmania were suitable for any *Salmonidae*, they would already exist. Dr John Gray, keeper of zoology at the British Museum, told a British Association meeting propagation was 'an unnatural mode of proceeding, and such is not practiced in the cultivation of any other animal. I cannot see any practical advantage that can possibly be derived from it.'

More annoyingly to Youl, Gray took a personal swipe: large sums had been spent trying to place salmon in Australian rivers 'but the many

Dr John Gray: 'nothing more absurd than introducing salmon to Australia'. Photograph ©Wellcome Collection.

failures show how little those who undertook the task were acquainted with the most common physiological questions connected with the removal of fish, and how small was their knowledge of the habits and peculiarities of these fish which they which they proposed to remove.'

In Gray's mind, nothing 'could be more absurd' than attempting to introduce salmon. 'I have no great faith in the success of the introduction of the salmon into Australia.'

Dr Albert Gunther, the rising ichthyologist, sent a copy of Gray's paper to *Field* because, he said, it reflected the views of 'every professed naturalist' he had ever spoken to.

Appreciating Dr Gunther might become the arbiter of any salmon caught in Tasmania, Youl had his enthusiastic assistant Thomas Johnson respond. He said the comments would have 'astounded' and 'greatly perplexed' all those who reported artificial breeding successes. They included William Ffennell, the chief commissioner of British fisheries; Thomas Brady, head of the Irish Fisheries Commission; Robert Buist of Stormontfield; Thomas Ashworth of the Galway fishery; Frank Buckland, Dr John Davy and Robert Ramsbottom.

Lincolnshire, Youl's third attempt. Photograph courtesy State Library of South Australia.

Artificial breeding might not be 'natural', Johnson said, but that did not detract from the 'unmistakable value' gained at Stormontfield and Galway, and he hoped such facts would set aside 'crude fossil notions'.

In Tasmania, salmon notions were divided. For some, even seeing a hatchling was believing, while others wanted to see a grown fish in the hand, and for others a tasting was ultimate proof of the pudding. Youl knew ultimate judgment would only come with evidence that eggs had hatched, grown to migrate to sea, returned 'home' to breed, and been categorically and scientifically identified as salmon. That would be the unambiguous proof of salmon being sustained in the Southern Hemisphere.

As *Field* put it, Youl had 'nobly vanquished' the transportation challenge, and the rivers of Tasmania looked desirable, but 'it remains to be seen whether the salmon is of the same mind'. So, for his fourth mission Youl determined to have his biggest 'profusion of eggs', wanting 150,000 ova to be transported on another Money Wigram clipper, the 1100-ton *Lincolnshire*, due to depart in January 1866.

Meanwhile, William Ramsbottom twice feared the *Norfolk* experiment was in peril. He was 'distressed' one day in October 1864 to discover water escaping from one of the outlet pipes running below the salmon pond into the Plenty. Uncertain how many five-month-old fry had found an escape route, he placed a box at the pond outlet to capture any more seeking liberation while he spent nearly three weeks digging trenches nine feet deep to locate and repair the leak. He captured 240 fry in the box, but estimated half the of the 3000 fry admitted to the pond had escaped.

Five months later he was alarmed when he found several dead parr in the pond. He first suspected 'foul play', as he had regarded the small fish, 'as bright and as healthy as any I have caught ... at the Ribble'. When the death toll rose daily to 20, 30, 40 then 50, with the cause unknown, he decided to have a good number 'liberated without delay'. He drained the pond to a foot in depth to collect and free 419 parr, each measuring about six inches long, 'lively and frisky as any fish need to be' into the Plenty. He could only hope the fry would change their parr markings for a silvery coat for camouflage, make their way to sea as a smolt, and return one day as grilse to spawn.

Youl was shocked when London newspapers reported the liberated salmon – the first to be formally released into southern waters – to be the only known survivors of the whole *Norfolk* consignment after a 'sudden and unanticipated mortality'. This followed a report in the Melbourne *Age* that 'a reliable private source' advised, contrary to Tasmanian reports that only a portion of surviving salmon had escaped, that the liberated parr were 'as far as can be ... ascertained, all the living representatives of the great salmon family at present inhabiting the waters of Tasmania.'

Field said the report underscored that no one could expect much beyond 'useful experience' after learning how to ship a fair number of ova for the first time. 'When they have them there their troubles are only beginning.'

Robert Officer quickly assured Youl the troubles were exaggerated. He believed the escaped and liberated parr would be 'not less than 2000 *finding* their way into the Plenty'. And the accidental escape might be seen as fortunate, as Francis Francis and Frank Buckland believed in their early release.

Officer relayed Ramsbottom's feeling that 'no apprehension need be entertained of any seriously injurious results as regards the ultimate acclimatisation of the salmon'. In fact, the *Mercury* reported the 'startling' news that he now estimated the salmon numbered 'not less than 6000, and ... may be as large as 10,000', while his trout estimate had risen to perhaps 400.

But Youl was troubled that at least seven young trout were known to have long been in the salmon pond. Long convinced trout would 'gobble up all my little salmon', he was not comforted by Salmon Commissioners advising 'none of us think that any of the salmon were injured by the trout; they were seen feeding and consorting together in perfect harmony on all occasions.'

Potential cannibalism did not sit with 'perfect harmony' for Youl, who acidly told *Field* that 'I will only just observe that trout, having been permitted in the same pond with the salmon, grow so much faster that I have very little doubt they have destroyed many of the smaller salmon fry'.

There was some welcome harmony in the Ramsbottom household. A week before Christmas 1864, 30-year-old Elizabeth, with five-year-old Jane and six-month-old Robert, boarded *Great Britain* in Liverpool. The captain ensured passengers and crew enjoyed 'a jolly Christmas' with geese, pork, tarts, plum pudding and port and sherry. It was less jolly when they arrived in Melbourne, where the government seized a United States Confederate war steamer, *Shenandoah*, amid fears the colonies were being drawn into the ripples of the Civil War. It was a relief when the Confederate vessel finally left, with some new colonial recruits, ultimately to fire the last shot in the Civil War.

In February 1865, Ramsbottom welcomed Elizabeth to a new cottage beside the Salmon Ponds. The small, single-storey building of pit-sawn weatherboards and hand-made convict nails, with a shingled roof, two chimneys and a veranda, sat amid a pioneering use of grass and Royal Society plantings of hundreds of trees and plants, some transplanted from Japan, China and Mediterranean countries. In an amphitheatre of hills and fields, the setting was described as 'Paradise' for man and fish.

The couple had endured long separations since their

Ramsbottom and family at cottage, Salmon Ponds.

wedding, with Elizabeth not joining William in Melbourne for 14 months and then enjoying only brief times together between the *Beautiful Star* and *Norfolk* voyages. After arriving, Elizabeth was soon pregnant with another son, Henry James.

Two months later, in April 1865, with the world reeling from 'Mr Lincoln is Dead!' news from America, Ramsbottom laid another historic imprint: after hatching around 300 of the 'unwanted' brown trout from the *Norfolk* shipment, he released about 40 fry into the Plenty: the first into Southern Hemisphere waters. The rest he retained as breeding stock, the genesis of the family tree of all brown trout across the island, the Australian mainland and New Zealand.

And concerned about suggestions the Tasmanian troubles were just beginning, Ramsbottom wrote to Francis Francis saying 'no very dark cloud hangs over our future prospects'. In the letter, which Francis had published in *Field*, Ramsbottom recalled how his father famously released 600 marked smolts from the pioneering Doohulla hatchery in Ireland and more than 100 returned as grilse. Based on that result, and with nearly 500 healthy parr released into the Plenty, 'we have every reason to look forward very hopefully to see a nice little breeding stock here', as while Antipodean enemies were unknown 'we have no cause to think that their risks will be greater here than on the Irish coast'.

As for the Francis and Buckland trout ova gifts, 'if you could only just see our trout now. I am persuaded you would say they were worth all the trouble and money that have been expended on all the experiments.' He had between 200 and 300 trout, under 16 months, but already 10 inches (25 centimetres) long despite living through two winters when they did not grow much.

Francis was pleased with the trout news, but doubtful the salmon number was sufficient to stock a river. However, 'if by good luck only one or two pairs of grilse should succeed in breeding ... the experiment will be pretty safe.'

To meet the desire for more salmon ova, in the 1865 winter Youl directed Ramsbottom Sr to fish the Ribble and Hodder near his Clitheroe home; his son Westall the Itchen and tributaries near

Southampton; Thomas Johnson the Tyne and Tweed; Frederick Allies the Severn and Teme.

But the salmon trail was again beset by snowstorms and heavy floods. Swollen, muddy rivers meant the salmon gatherers could not use nets or see any fish in their spawning beds. It was 'a hopeless task', they complained, but Youl strenuously urged them to persevere. He urged one, two or even five-pound rewards for anyone who caught a spawning fish.

The Ramsbottoms and Johnson finally delivered about 87,000 salmon ova and about 10,000 sea trout, and Allies about 500 trout ova. The tally was disappointingly short of Youl's 150,000 salmon target, but he hoped to still get another 30,000 before *Lincolnshire* departed.

The sea trout would cause no end of trouble in the salmon quest. They were brown trout which hatched in fresh water but chose to move to the sea before returning to spawn. The 'anadromous' fish had different regional names, such as peal in England, herling in Scotland, white trout in Ireland, and sewin in Wales. In Tasmania, where for many the visual differences between young salmon and sea trout were marginal, the more magical sounding term 'salmon trout' was persistently adopted by the Salmon Commission, Morton Allport and newspapers. The Commission saw the fish as 'an acquisition especially valuable ... these fish nearly approach the true salmon in size ... as well as in their qualities as an article of food.'

When it was clear no more salmon ova would be forthcoming for Youl, 'to my great grief', he was back on his dockside cushion in another freezing winter, focused on the lessons of ensuring healthy moss and light packing. 'I (took) all the pains I could to do this and would not trust anyone else.' He packed 120 boxes of salmon, 20 sea trout, and one brown trout. His fingers were again numb and bloody from broken ice.

He filled the remaining icehouse space with other experiments: half a dozen fresh-laid hen's eggs 'to see if they will hatch in Melbourne', along with six apple trees, two boxes of bulbs, two bundles of Scotch heather for 'grouse ... soon (to) be added to salmon', and two boxes of silkworm cocoons gifted by Paris acclimatisers.

On Saturday 20 January 1866, almost two years to the day since *Norfolk*'s historic departure, Youl wished *Lincolnshire* Captain Edward Charleton good sailing in delivering his '1 tank ova' to a waiting Ramsbottom in Melbourne. There was still cause to be anxious, as just a week before Britain was distressed when *Lincolnshire's* sister ship, *London*, loaded with emigrants to Australia, tried to make a run back to Plymouth in 'mountainous high seas' before it went down with 239 souls lost. The nation shed tears over distressing messages found in bottles washed ashore: 'the ship is sinking ... may we get home! ... no hope of being saved ... farewell dear wife and children, may God bless you all ... God bless my poor orphans.'

So Youl felt 'much anxiety' about his 'babies' when he heard *Lincolnshire*, at anchor off Greenhithe on the Kentish coast, had been hit by the brig *Spartan*. It forced Captain Charleton to return to the East India Dock for repairs. Youl feared 'a disaster' as it risked the ova being imprisoned in the icehouse for more than 100 days, the life-death limit, before it reached Tasmania.

Lincolnshire left again on 25 January, but hurricane winds meant nine days were lost in 'the chops of the channel'. Conditions improved for the remainder of the voyage, but 100 days had passed when *Lincolnshire* arrived at Melbourne late at night on 30 April. Ramsbottom immediately clambered aboard to check the precious cargo. Inside the sealed icehouse he saw enough to calm his concerns, and was back on deck at 6am to supervise the ova transfer to the warship steamer *Victoria* for the final leg to Hobart.

The *Times* correspondent described 'our Salmonians' out in force, solemnly watching 'as great officers of the State may assemble in an anteroom on the birth of a Crown Prince'. The loading task took 12 hours, and with ice blocks frozen together, Ramsbottom could only grimace as men needed iron bars to free the small ova boxes so they could be encased amid 15 tons of ice in larger containers.

Victorians were happy this time for all salmon eggs to go to Tasmania to maximise the chances of success and await their share of fry in due course. They retained the single box of brown trout ova, destined for the estate of William Clarke, perhaps the richest man in Australia and soon to earn fame for hosting an event which birthed the cricket Ashes.

But the box was a coffin, all the ova 'quite dead'. Some feared the salmon shipment also 'the ghost of a departing idea', but *Victoria* Captain William Norman was determined not to deliver dead salmon. Concerned by steam engine vibration, he kept *Victoria* at half-power, which with dense fog meant a 40-hour journey became a tense three days.

But Tasmania sensed a coronation. Five months earlier, the *Mercury*'s Christmas tidings declared 'a salmon cutlet of Tasmanian growth is the very least thing we promise ourselves for next Boxing Day'. With escaped and liberated salmon somewhere beyond the Plenty, and another shipment about to dock, the population could now 'think of the delicacy in store for you in the shape of salmon cutlets, and what an agreeable change it would be from beef and mutton when the thermometer stands at 100 degrees in the shade'.

It was late at night in New Norfolk when the latest 'little strangers' arrived, but the township stayed up to provide a skyrocket welcome. The cargo was again towed upstream before patients from the lunatic asylum helped carry 16 heavy crates of ova boxes. Those making two journeys earned 15 shillings, some bread and ale.

Great fears arose when the first few boxes revealed nearly all the ova to have perished. But the results progressively became 'magnificent ... they lay like pink pearls, and sure no pearls ever promised to benefit the human race one hundredth part so well'.

After six days carefully picking out dead ova, Ramsbottom felt he had some 50,000 salmon eggs,, a much better result than with *Norfolk*, and 7000 sea trout in healthy condition.

The *Argus* reporter marvelled: 'Here, in an obscure corner of a small remote English society, was being worked out one of the most remarkable experiments of the age. We were writing an invaluable page in natural history. In faraway England, scientific men devoted themselves to efforts the fruition of which they would never see, while we were completing that which had been begun by other hands 16,000 miles away.'

The occasion, he said, was something 'eagerly looked for by every devotee of science in the world.' His dispatch to the world was unambiguous: 'This is to announce that the last and most carefully planned attempt to bring salmon ova from the best English streams to be acclimatised in the rivers of the Antipodes has proved the most successful of all. The reward of so much trouble, thought and preparation has been proportionally brilliant, and Australasia may now count upon having her salmon, in a year or two, by many tens of thousands, for the chief difficulties and the greatest dangers have been overcome.'

But in the wake of Darwin's observations about human, landscape and species adaption to European colonisation, no one could know if salmon could meet the greatest adaption challenge. But Ramsbottom sensed, or hoped, that some had adapted since hatching and were preparing to return from the sea to produce Plenty offspring. 'I think I can safely say now that all that were liberated last March are by this time in the briny ocean.' If they survived and found their way back to their nursery water to spawn it would be the 'crowning test' of success.

In London, salmon was also of hope and hype. The *Times* excitedly reported 'a remarkably fine salmon' of nearly 13 pounds (5 kilograms) caught near Greenwich. Once one of the great salmon rivers, the Thames had not seen a captured salmon for at least 30 years, having become a giant sewer fed by the waste of industry and two million Londoners. An excited Frank Buckland, now Fish Culturist to Queen Victoria, told William Ffennell, inspector of fisheries, the fish could be one returning from the sea after being artificially propagated in 1861. But Ffennell said 'I cannot believe it possible for a salmon smolt to descend with life through the extent of filth which he must now encounter.' The *Standard* suggested the catch was more likely a Medway salmon simply passing across the Thames mouth, as any salmon endeavouring to swim up the Thames would 'not find it possible to dine at Greenwich without any disturbance to his stomach'.

But Buckland remained optimistic. Having already put lake trout ova into the royal ponds at Windsor, he predicted that within five years he would also have the honour of laying a salmon on the Queen's table.

Tasmanians anticipated trout and salmon on their tables well before then. Trout had been released and others were beginning to breed in their pond. Salmon had escaped or been liberated into the Plenty, while those retained for breeding were in good health, delighting pond visitors with their silvery bellies flashing on the surface, almost as if trained. One considered Ramsbottom had 'in a remarkable way ... retained the confidence of his finny proteges. They come to be fed on gentles (maggots) and house flies, whenever invited, and seem to enjoy the scramble as much as the food.'

Ramsbottom struggled to count how many he had. 'The only way I can make sure of seeing any is at night when I walk around the pond with a candle (to) find the little fellows scattered about close to the edge.' He was pleased one night in October to see 14 young salmon in silvery migratory dress, signalling their readiness to head for the sea, 'swimming round and round in the pond looking for some place by which they can make their escape ... it seems a great pity to keep them in confinement, but we must do so until we know the result of those already away at sea.'

The salmon dream was very much alive, especially with Ramsbottom's prediction, or hope, that the escapees and liberated would announce their return around January or February 1867.

17

PATIENCE & IMPATIENCE

On the second anniversary of the arrival of the first salmon and trout ova, Tasmania was impatient to see the first salmon returning from the Tasman Sea. Further eastward, New Zealanders desperately wanted their own.

As with Tasmania, settlers from Scotland wanted a 'home away from home ... a finer Scotland' in new places they called Dunedin, Avon, Roseneath, Ben Nevis, Bannockburn and Invercargill. They too wanted all the 'good' from home and saw their rivers and streams as especially well-suited for salmon and trout.

Pioneering sheep farmer and naturalist Mark Stoddart declared: 'We feel a want ... the song of the bird, the consciousness that deer may be feeding on the slopes of our glittering alps, or the salmon and trout splashing in our clear streams.' On introducing salmon, his father, angler-author Thomas Tod Stoddart, felt it 'will eventually be accomplished' by using Youl's method of ice retarding ova growth, but 'it would not bear packing in moss'. He also thought Californian ova a better option for New Zealand.

Youl believed New Zealand could, and should, have a share of his salmon, and urged New Zealand's agent in London, John Morrison, to secure a working model of the Stormontfield hatchery for the Southland Government to ready for any ova.

Andrew Johnson, a Birmingham farmer and fish-culture hobbyist who saw New Zealand as his new life, did not believe in the hope that thousands of shipped ova might survive. He favoured a few grown fish, artificially propagated and protected, as the source of 'acclimatised' eggs and fish, and left England with an assortment of English fish, from minnow to salmon. After they all died from lead poisoning, he became secretary-curator to the new Canterbury Acclimatisation Society, which with Nelson, Christchurch and Dunedin societies wanted to 'participate at the earliest possible moment' in the acquisition of salmon, with Tasmania's help.

But Tasmania felt it too soon to share any of its salmon ova bounty. Trout, however, was in play. Advertisements urged those in New Zealand, Victoria and New South Wales anxious to populate their rivers and streams to 'lose no time' registering their interest in obtaining trout ova from the Plenty Ponds.

Suffering a persistent chest cough, Ramsbottom spent the 1866 winter stripping some of the 133 trout he had retained in captivity. He proudly said the two-year-olds, some weighing more than a pound (half a kilogram), exceeded the standard usually reached at the same age in Great Britain. He secured about 4000 fecundated ova, some of which hatched, astonishing British scientists and sceptics who ridiculed the notion that fish with any migratory instinct could breed without going to sea.

Soon after, he was thrilled to see five pairs of the original 38 released trout had created spawn redds in the Plenty. He caught one male of 22 inches (56 centimetres), estimating it would have weighed between five and six pounds.

Of all the pink-coloured territories on the map of an Empire 'where the sun never sets', Tasmania was the first and only one to have pink-coloured fish outside Britain. The last-minute, unwanted trout gift on *Norfolk* was now seen as beyond all risk of failure. Officer told Youl 'my dear sir, the establishment of trout in our rivers is an established fact, no longer to be doubted by the most incredulous'. It was a 'surprising … unprecedented' achievement, especially when it had failed in a more vaunted colony. In the mountain streams around Madras, India, Youl supporter Dr Francis Day had striven to transplant trout ova but water temperature, insects and dust killed the prospects.

Reaching the time when Tasmania could share some trout bounty with those who had lent support, Ramsbottom packed about 1800 brown trout ova into two boxes of wet moss, to be chaperoned to Victoria by William Sangwell, the Government House head gardener who helped build the Plenty Ponds. He carefully nursed the boxes on the *Southern Cross* to Melbourne, then by train north-west to Sunbury and the estate of William Clarke. While Victoria had lost confidence in ever seeing any of the earlier transplanted *Norfolk* salmon, the *Australasian* said the expense and effort would not be in vain as the brown trout and 'salmon trout' seemed assured, although many would not believe acclimatisation 'was an actual fact until they see it smoking on their plate'.

Thoughts of smoked trout dissipated eight weeks later when Clarke's Sunbury stream flooded. He 'feared all the young fish had been destroyed', but a few weeks later a farmer reported 'trout' in nearby Keilor Creek, which the Victorian Acclimatisation Society accepted as 'evidence' not all had perished.

Ramsbottom next turned his attention to the north of Tasmania, which had been frustrated its rivers, with English names like Tamar, Mersey, Forth and Esk, had not been populated with any of the 'great blessing' delivered by Youl, one of their own. Ramsbottom packed about 800 spawn into alternate containers: a box of damp moss, and his father's technique of a bottle of water, tied by string to his finger to minimise vibration on train journeys. But the bottle approach was useless on a rollicking coach – by the time he reached Bridgewater, barely 12 miles

(19 kilometres) away, it was clear those eggs would not survive the remaining 128 miles (205 kilometres), so he transferred them into the box of moss.

In Launceston, Ramsbottom deposited 420 surviving ova into hatching boxes in the garden of Charles McArthur, a successful agent for the Liverpool and London Insurance Company. McArthur's plan was to nurse the ova until they became fry, then place them for 12 months in a park fountain opposite the church Youl's father had pioneered. Searching for streams which might be suitable for salmon or trout 'when the proper time arrives', Ramsbottom declared a pond on the Strathmore estate, owned by Youl's in-law Cox family, 'the best place he had ever seen'.

In the south, eyes anxiously scanned for anything resembling the grand prize of salmon returning from the sea. The Salmon Commission did not see the immediate absence of salmon evidence as evidence of absence. 'The Commissioners cannot take it upon themselves to say that none [have] returned ... a thousand fish in such a stream as the Derwent might pass and repass without attracting notice.'

Sceptics saw hype rather than hope, but the latter was resilient. As Anthony Trollope would soon observe, salmon was the 'great desiderata' to address 'the feeling of disgrace and the aspiration for a different state of things'. The disgrace was 'evil' convictism and the tragic demise of the Aboriginal population. The aspiration was to re-imagine Van Diemen's Land as a Tasmanian paradise standing on its own economic feet, with a salmon industry and anglers and tourists offsetting a human and capital exodus to Victorian and Californian gold rushes, and declining exports to economically straitened Britain, and Civil War-ravaged America.

Even outsiders hoped for salmon salvation. The Melbourne *Argus* said Tasmania needed to be more than the country's 'impoverished tributary'. If not, then once Bass Strait could be crossed within 12 hours, 'as it will be eventually', it would descend into a resort for 'invalids and tourists'.

Youl's hopes rose. Waters had saved the island before. Seafood and fish had helped first settlers avoid starvation. Then whales and seals became a major economic contributor, with three thousand southern right whales taken for sperm oil and fur in a single year before greed took its inevitable toll. Then oysters became the most valuable primary industry with an unsustainable 20,000 million exported before it also faded.

hen he received a telegram in July 1866. The medium was as remarkable as the message: it came via a series of ships and telegraph operators in Ceylon (now Sri Lanka), Bombay, Alenxandra and Paris; a steamer to the London docks; train to London and finally a messenger's satchel bag to his front door at Clapham Park.

Youl rushed to his study to quill a note to the *Times*: 'I have this moment received a telegram from Melbourne announcing the safe arrival of the *Lincolnshire*, with upwards of 100,000 salmon, sea and brown trout ova, and bringing also the good news that forty per cent of the whole number was hatching in the Plenty ponds.' He also had the report from the Salmon Commission that at least 2000 *Norfolk* salmon fry between nine and 12 inches had gone to sea.

Youl hoped the next time he wrote to the *Times* it would be to reveal the salmon had returned, as 'until this takes place, we cannot assume that the experiment has been a complete success'.

He was pleased his apple trees, bulbs and heather plants had arrived safely in Melbourne, but the hen's eggs, mistakenly sent to Tasmania, were another story, thanks to Morton Allport who sniffed that 'whoever sent (hen's eggs) must have a very limited knowledge of natural history'. To Youl 'this reminds me forcible [*sic*] of the ridicule I was subjected to for many years for expressing an opinion that the eggs of salmon could be sent to the Antipodes and hatched there.' He explained the hen eggs experiment was suggested by Dr Day, who in a lecture to the Royal Geographic Society of London said he wanted to test whether bird eggs, including grouse and partridge, might be successfully transported.

'I very much wish Mr Allport – who expressed an opinion that the eggs looked fit for culinary purposes – had put his opinion to the test, and if one of them boiled for his breakfast and it turned out as good as it looked, would have placed the others under a hen. We should then have had some knowledge as the feasibility of sending eggs long distances in an icehouse on board ship.'

Allport also irritated Youl by concluding that different success rates and discrepancies with moss and ova in boxes variously signed by Youl, Robert Ramsbottom and Thomas Johnston were due to different individual packing skills. Youl took exception, writing to newspapers that Allport was 'totally erroneous', as only he and Ramsbottom had packed the eggs, each alternately placing moss and ova in a consistent pattern.

Despite the failures of *Curling* and *Beautiful Star*, of which Allport had been scathing, Youl said he was still confident of success, and pointed out if he had not eschewed the opinion and ridicule of others rather than his own 'well considered plan ... (then) in all probability you would now not have either salmon or trout disporting themselves in your rivers.' For his part, Allport maintained the more significant impact had been his manoeuvring to have trout introduced against Youl's wishes. This, he wrote, had been pivotal to maintaining faith in salmon, and funding for the Salmon Ponds, without which the whole quest might have foundered.

A *Times* correspondent said success from salmon ova was deserved 'for ... hearts, stomachs and pockets have been in this business for years'. The newspaper had moved from scepticism about propagation to declaring 'the seed of salmon [may be] deposited in boxes like peas in a garden', but in the garden of Tasmania 'we can hardly believe that salmon are perfectly naturalised' before the 'crowning result' of evidence that salmon had returned from the sea.

While waiting for any sighting, Ramsbottom estimated that of the 45,000 salmon ova on *Lincolnshire*, perhaps 30,000 had survived the voyage. By the end of hatching, his estimate was down to about 6000 fry, along with 1000 sea trout and an unknown number of brown trout. It was not a high percentage, but Ramsbottom thought it 'very encouraging [when] the ova had been packed up in such an unnatural way for 112 days with the tossing and jolting they must have had'. With all the uncertainties of fertility, fecundation and hatching, the Scots had long suggested as few as one in 6000 salmon ova survived to become a grown fish, let alone return to spawn. And this was without the unknowns of extended retardation across the world into alien waters.

The Salmon Commission admitted the *Lincolnshire* result 'scarcely comes up to expectations' but rated it a 'great success', with hatching rates not far short of those attained at the famous Huningue breeding facility in France. Not everyone was so content. The Victorian Acclimatisation Society expressed 'a great ... surprise and disappointment', and a Melbourne writer, calling himself 'Minnow', questioned a 'mysterious disappearance ... where are the others?'

His criticism of Commissioners for 'mismanagement' in raising public expectations to 'ridiculous' levels irked Officer and Allport. They protested that the Commission's exertions 'have been made the subject of miserable and illiberal comment' by those unaware of the difficulties involved, when the success to date 'should have won for them the gratitude of the public at large'. Youl might have enjoyed their discomfort.

To settle the 'missing' salmon controversy, the *Australasian* referred the matter to 'a high authority at home', Francis Francis. The angling editor was not surprised by the losses: 'If you consider what that ova had to go through it is only too wonderful that you do contrive to hatch and rear any at all.' He also urged more faith in the sea trout ova he had sourced for Youl 'from a friend who owns the salmon fishery on the Itchen in Hampshire ... they are perhaps the finest breed of salmon trout in the world ... almost as valuable as salmon.' They were good eating and good sport, taking bait and fly better than salmon.

Francis found the progress of trout and 'salmon trout' reassuring, but recognised salmon love was unrequited. 'O that we could hear of half a dozen grilse having been seen ploughing their way up the maiden waters of the Derwent! That would be news worth hearing. Let us trust that the good time is really coming and then in fish culture nothing need be held impossible.'

Francis Francis, from *A Book on Angling*, Longmans, Green, and Co. London. 1876.

That good time had been coming since the first *Norfolk* hatchings in 1864, but no one could know if salmon had even made it to sea, survived, or could find their way back to an adopted 'home'. Thousands of pages had been written on salmon, but much remained a mystery. How did it transition from fresh to salt water? Where do the fish go when at sea? How do they know when to return to spawn? How do they find their way back? As a contributor observed in London's *New Sporting Magazine*: 'We are still in the dark.'

It hadn't stopped Tasmania becoming 'Salmoniana'. In a popular pamphlet of that name, Scottish journalist Thomas Just, using the pseudonym Barri Couta, and pioneering cartoonist John Manly, as A Trumpeter, published witty verse, cartoons and caricatures. In the saga of 'this queer home mania' of acclimatisation, they said, the 'queerest' was the mania for salmon. It referenced that a few years before in London: 'A cove who profess'd one or two things to know/ Found out that the eggs of salmon so nice/ Could be kept an indefinite period in ice.' This led to local savants being 'all agog' and while the administration in Hobart Town was normally 'as spry as a log ... right off the reel the colonial parliament/ made a vote of some thousands to try the experiment!'

It concluded that if the fish at sea were faring well, then 'Tasmania's attempt will in history's page/ Be described as the greatest success of the age!'

Youl's hope of history's page was warmed by more people declaring 'Bravo!' to his exploits, his packing solution described as 'ingenious' and talk of 'fool' faded. From the scientific capital of the world, he was invited to Paris to receive a silver medal from the Societe Imperiale Zoologique D'Acclimitation for 'introduction en saumon en Australie'. And to Napoleon III's Exposition Internationale de Peche, staged at Boulogne-sur-Mer and Arcachon near Bordeaux, designed to show Europe the 'science of the sea'. He received a medal from La Société Scientifique d'Arcachon for sending a paper 'on the means used to transport salmon eggs to Australia' along with a sample ova box.

Pleased as he was, it irked Youl to be more readily, and substantially, recognised in Europe than by his own countrymen. He made sure the

Tasmanian newspapers knew of his French awards, moving Hobart's main newspaper, which once said the quest could never succeed in his hands, to again eat humble fish pie. 'Mr Youl has achieved a great piscicultural triumph,' the *Mercury* proclaimed. 'All of us know that Mr Youl has not always been successful in his experiments. Failures are never forgotten, and it is extremely seldom that fair and full allowance is made for them. Benefits, on the other hand, come as the good gifts of Providence.'

The failures were steps to success, and determination and energy had seen Youl 'surmount all difficulties' to realise 'the crown of final victory'. But having been conferred an 'immense benefit ... commercial value ... recreation and gastronomic luxury', the legislators and Salmon Commission had failed to recognise and reward Youl accordingly.

'Mr Youl has done his duty to the colony; it remains to be seen how the colony will do its duty as regards Mr Youl. If a community did not publicly recognise those whose efforts to benefit their country had no reward beyond self-satisfaction it was guilty of gross neglect and ingratitude.'

Guilty as charged, Youl must have thought. If only a returning adult salmon could lock in his recognition as the 'Mr So and So' who delivered the legacy of salmon. Robert Buist, the respected Tay fishery superintendent, who had converted from propagation 'sceptic' to 'ardent disciple of the science of pisciculture'. He had witnessed salmon eggs 'deposited in our boxes like peas sown in a garden, and I saw it come to life ... man may successfully cultivate the waters as he cultivates the land,' but, he said if the newly developed smolts in Tasmania returned as grilse 'the salmon will have passed the Rubicon'.

Could someone in Tasmania prove the Rubicon had been crossed? Anticipation was fired by the son of a British soldier. Amid spring showers of October 1866, Henry Bayly and some teenage friends had their rods in the New Town Rivulet running from the brooding Wellington range overlooking Hobart down to the Derwent near the Queen's Asylum for orphaned, destitute and neglected children. The boys hoped to catch a blackfish or an eel, Tasmania's only two large native freshwater catches.

What Bayly landed was only eight inches (20 centimetres) long and unfamiliar, but Scottish-educated asylum surgeon-superintendent, Dr

John Coverdale, sensed some significance. He forwarded the fish and a statement about its capture to the Salmon Commissioners, who were convinced that after hatching the fish had swum 30 miles (48 kilometres) from the fresh water of the Derwent's upper reaches to the salty water around Hobart. No one could fathom the 'apparent eccentricity of this young traveller' and why it had not gone to sea to fatten to 10 times its size, or whether it merely ventured up the rivulet to seek fresh water or escape some enemy.

But the Commissioners saw nothing curious about its identity. It was what they were looking for. They declared it 'a fine, healthy smolt'. Officer saluted 'this famous fish' and Allport proudly carried it in a preservation jar to an intercolonial exhibition in Melbourne. Labelled a 'salmon [two-year-old] smolt', it had the *Argus* newspaper enthusing about 'a real Tasmanian salmon ... as fine a fish of its age as ever was seen in the United Kingdom'.

Allport happily received natural history medal, and the acclaim of the Acclimatisation Society of Victoria for the 'gratifying proof'.

But ultimate proof remained an adult salmon returning to spawn. When Youl heard of Ramsbottom's landing of a large brown trout, he thought it augured well for salmon. And when Officer told him of a new generation of trout hatching, he immediately wrote to *Field* that 'they are the first *Salmonidae* ever hatched in the Southern Hemisphere, and I think their success augurs well for that of the *Salmo salar*'.

He hoped the trout news would quieten the sceptics, although Francis cautioned that any trout success was not necessarily a guarantee of salmon success. Still, Youl much enjoyed the 26 January 1867 anniversary dinner commemorating the foundation of 'the first of the Australian colonies', with *Field* publishing his proud note, and the *Illustrated London News* featuring a woodcut illustration of the Salmon Ponds, declaring it would address 'the want of a good eatable fish in most of the rivers and streams of Australia, Tasmania and New Zealand ... a serious drawback in the comforts and enjoyment of the colonists.' The leading angling journal and the world's first illustrated newspaper gave him something to dine out on.

The salmon chase was the talk of the day on both sides of the world. Some, such as William Ffennell of the Salmon Fisheries Office, still believed the artificial rearing of salmon was a 'comparatively useless ... unproven (interference) with nature', but the *Times* was dismayed England had 'killed our goose for its golden eggs', such that 'even distant Norway sends us more salmon than we can catch ourselves, though the shores of this island can show the finest natural salmon grounds in the world. Scotland and Ireland are far ahead of us in this kind of farming. We cannot afford to neglect any source of wholesome and palatable food.'

Youl anticipated the universal reaction that would come with evidence that his 'little Antipodean samlets', given to the care of others with 'as much interest and anxiety as if they were my own children', had grown up and were ready to rear another generation.

It was time for salmon to break their silence and show themselves. For William Ramsbottom – who had proclaimed February 1867 as the likely time for the first salmon to return from the sea – it was time to begin a salmon vigil.

18

VIGIL & VIRTUE

Ramsbottom spent the southern summer walking two miles every morning and evening to and from his cottage to Dry Creek, a spot downstream from the Plenty-Derwent junction. Still, deep water ran for about 500 yards (457 metres), with rapids beyond the reach of the tide and a gravelly shallow at the upper end.

In this perfect setting for salmon seeking a return to their nursery to spawn, he hoped to be finally rewarded for the midwifery efforts he described as 'trouble' and 'monotonous'.

When two New Norfolk women 'of the highest respectability ... worthy of every trust', and Captain Joseph Oakley, a former whaler and owner of the Bush Hotel, reported 'large and strange fish, never before observed by them, seen to leap in the Derwent' Ramsbottom supposed they were returning salmon.

But he could not be sure if the fish they saw were of grilse size, or whether they could be 17-inch (43-centimetre) trout such as he had fished from the Plenty, and all Waltonians knew trout could make as good a show in the water as any salmon. So he cautiously refrained from drawing positive conclusions.

On his vigil he was often accompanied by the water bailiff, James Lumsden, described as 'an old salmon-fisher from Scotland', familiar with Spey salmon at Elgin, Scotland. The pair spent six hours a day on the Derwent banks, only to be disappointed six hours a day. Ramsbottom was unsurprised: 'This need not be wondered at when it is taken into consideration what a length of river they have to roam about in.'

Of the salmon's ultimate return, he had 'no doubt', but others insisted the salmon would never be seen again. They would never return due to a shortage of food, unfamiliar disease, or unsuitable water temperatures. Or had been taken by predatory trout, some of the nine sharks known in Tasmanian waters, humpback or right whales, fur seals, or barracouta. Or had their migratory antenna rent asunder in a foreign hemisphere.

But on St Valentine's Day 1867, men of God delivered blessed news. Dr Daniel Murphy, the new Catholic bishop of Hobart, his vicar-general William Dunne and Dr James Hayes, Dean of Melbourne, had made their own pilgrimage to the ponds. Returning to Hobart, they saw movement in the river near Dry Creek. Jumping out of their carriage, Dunne and Hayes saw a large fish twice leap from the water above the rapids, before gliding under the surface. Dunne, who in his younger days had lived near a Shannon fishery, at once pronounced it salmon.

God be praised! The news quickly raced to Hobart. The *Mercury* said Ramsbottom had long predicted this was the time salmon would return, but while large fish had been seen leaping 'they were not seen by persons who could identify them'. But Rev Dunne's judgment meant 'the thing appears to be put beyond all doubt'. It could be 'safely concluded the salmon are getting back from the sea'.

Ramsbottom was encouraged but still frustrated. It would have been 'more gratifying to me [and] perhaps more satisfactory ... had I seen the fish'. He felt quite certain it must have been a salmon 'for I believe Father

Dunne is about as familiar with the sight of a salmon as myself. He could not be mistaken for we have no fish so high up the river like salmon, either in size or color [sic].'

Notwithstanding his faith in Dunne, Ramsbottom wanted to see salmon with his own eyes. Only then would he pronounce its return 'certain'. But the Salmon Commission wasn't waiting. The numerous sightings by clergy and others were 'unquestionable evidence of the presence of the returned salmon'. Word spread quickly. A 'salmon has been seen in the Derwent—certain' telegram spread to other colonies and beyond, with the *Argus*' 'Summary for Europe' stating 'there seems to be no doubt ... a very important ... experiment in acclimatisation has ... been crowned with success.'

Morton Allport wrote, not to Youl, but his associate Thomas Johnson in London: 'I write at once ... as to the return of the fish from the sea. After carefully sifting all the evidence, no doubt remains in my mind that a number of our first shipment have indeed returned to the Derwent, and that the experiment has resulted in entire success.'

More sightings of leaping fish were reported by water bailiff Lumsden, and the *Mercury*'s debt collector, George Simonds, who grew up near the famed Frome River in Dorchester. Each judged them grown salmon.

The *Mercury* had heard enough: 'The return of some of the salmon from the sea is beyond all doubt ... all doubt as to the success of the experiment is now at an end.' So too the *Australasian*. Under an 'ADVANCE TASMANIA!' headline, it couched the salmon as symbolic of a new Tasmania after years of enduring evils imposed from home. There was nothing to prevent the colony capitalising on its climate 'perfection', splendid harbours, inexhaustible forests, and fishing waters superior to continental Australia.

Some locals, however, without discrediting 'those who believed they saw the salmon in the Derwent', felt that until Ramsbottom had a specimen in hand the public was entitled to assume no 'certain proof'.

Ramsbottom was becoming desperate for that proof. He became excited when he spotted one large fish just above the surface, accompanied by a whirl following its movements, but it wasn't enough for him to 'identify on oath' what he believed, or hoped, it to be. He bemoaned his misfortune at never being 'at the proper place at the right time'. But if he had to walk the banks all season 'and not have the good luck to see a single salmon, I shall nevertheless feel quite confident that the fish have returned from the sea.' He just hadn't seen one to fully satisfy himself.

After early breakfast on Friday 15 March 1867, he started off 'for the 69th time' to Dry Creek. At 10am his assistant John Stannard rushed to tell him half a dozen salmon had been seen playing in another section of the Derwent. 'The joyful news again put new life into me and I at once started off for the spot. I meant to see a salmon if one was to be seen at all during the day.'

After a hot two-hour trek, he sat on a log about 15 feet from the river and waited. And waited. After more than two hours, 'I had got really impressed with the idea that I was about to be the last person to see a salmon'.

Then, just after 3pm, he heard 'a kind of rush behind and rather below me'. His pulse quickened. 'I looked around and saw it was the motion of a large fish, but I could not persuade myself it was a salmon.' Frustratingly, the fish disappeared. Ramsbottom scanned fruitlessly for 20 minutes before 'I saw the partial rise of a fish, head and breast out of water'. But he was still not completely satisfied as it was a distance away and 'did not make the same splash as I thought a salmon ought to do'.

But his mind was racing. 'The fish looked larger than any I had before seen in these waters so that I now got very uneasy in mind, sometimes doubting, and at other times believing, they must really be the salmon.'

Then at 3.40pm his 'uneasy mind' was settled. 'To my exceeding joy I was delighted to see nothing else but a salmon jump clean out of the water and show himself broadside on, and judging from his appearance I should call him a grilse of about five pounds weight. On seeing the salmon, it is impossible for me to describe my joyful feelings.'

In 'considerable exultation' he raced to Officer, Salmon Commission Chairman Director, who immediately dispatched to the *Mercury*:

'OFFICIAL REPORT OF THE RETURN OF THE SALMON. This afternoon, between three and four o'clock, Mr Ramsbottom's long watching was rewarded by seeing two salmon leap from the waters of the Derwent, near the mouth of the Plenty, where they had previously been seen by various persons, whose reports are thus amply confirmed. Of the first of the fish thus seen by Mr Ramsbottom, the head and breast were distinctly visible, and in the second instance, which occurred soon after the first, the whole fish was exposed to view.'

Ramsbottom sent a note in the mail coach to Launceston. 'I have at last seen a salmon ... luck has come at last.' He saw 'a beautiful fish, I assure you, I should think from four pounds to six pounds weight. He showed twice, but the best the last time, as he rose about two feet from the water and broadside on; a better sight I could not have wished for.'

He had visited the river '70 times for that purpose and walked nearly 300 miles to and fro' but he was not complaining. 'I feel well repaid for the trouble and monotonous work ... [and] we have got the salmon back, and I think in goodly numbers too.'

The *Examiner* said that while numerous sightings made 'it scarcely possible to disbelieve', it had withheld 'entire credence' until it had 'the authority of Mr Ramsbottom himself'. Now the paper had it in his own hand. Under the headline 'SALMON, AND NO MISTAKE!' it declared 'any lingering doubt that may have existed ... can be given to the winds'.

It would be 'cheering news to transmit to the old country by the next mail'. And it was. Francis Francis said it was 'not at all improbable' they were salmon, as the leap of 'Master Silverside' returning to his native stream was not easily mistaken and he had no doubt Ramsbottom 'could tell if they were salmon a quarter of a mile off'.

Youl relished a report by the Melbourne correspondent of the *Times*: 'This mail takes to England news which will be grateful to naturalists and to scientific men generally.' After 'old colonists' laboured 'at great cost and with infinite pains ... the salmon is now naturalised in Australian waters'. After much dispute about the chances, 'the salmon himself [has] settled the controversy.'

While Francis hoped a capture would provide the final, undeniable proof – 'Mr Youl must look for every mail now with the greatest anxiety' – the *Times* and other newspapers across Britain hailed the experiment as 'the greatest triumph of this kind that has as yet been achieved'.

Queen Victoria and Prince Albert hoped an unprecedented world naval voyage, including five months in the Antipodes, might see their wayward Prince Alfred adopt a less debauched, more royal personage.

In manic competition among the colonies to demonstrate the greatest loyalty and affection for monarchy and Mother Country – some even called for Alfred to become the first King of Australia – Tasmania wanted to show it could not be bested for Englishness. Governor Thomas Gore Browne asserted it embodied the maxim of ancient Roman philosopher Quintius Horatius Flaccus: 'They who cross the seas change their climate, but not their feelings.' Tasmanians were 'Britons to the backbone', their loyalty and attachment not chilled by time or distance.

It was hoped Alfred, as president of the British Acclimatisation Society, could taste a fresh salmon cutlet. After decades of 'salt salmon in jars' and 'pickled salmon', Tasmanians relished the prospect of their own monarch of the stream on a plate before a royal prince.

In the leadup to the prince's arrival, the *Mercury* likened Youl to Christopher Columbus: it related the apocryphal tale of Columbus challenging sceptics of his new trade route ambitions by asking them to make an egg stand on its end. The egg toppled over several times before Columbus took the egg, cracked its top off, and 'lo behold!' stood the egg on its end. Youl had done 'even better by teaching us how myriads of eggs may be transmitted in a state of perfect preservation and be hatched 16,000 miles from the spot on which they were laid.' His discovery was 'world-wide in its beneficial influence'.

Not everyone was so effusive. A young English aristocrat, Charles Dilke, was on a world tour in his desire for Great Britain becoming 'Greater Britain' by embracing colonies including Australia and New Zealand. The 23-year-old found Tasmania 'the most English of all the Australian colonies', but disparaged the general 'melancholy', convict stain, 'indolence' of settlers and 'savagery' towards the local Aboriginal people. He suggested the only 'cure' for the 'Ireland of the south' might be annexation to Victoria.

He especially sneered at 'the salmon madness'. He wrote that his hosts at New Norfolk were fearful Rev Dunne, who had become 'the hero of the day' for swearing he had seen a true salmon, 'should be lynched if it were finally proved that he had been mistaken'.

Tasmanians had long endured being demeaned. But they would show Dilke and his ilk they were never to be underestimated. They would outdo other colonies and sceptics by setting a dainty dish before royalty. An HRH affirmation would be gold, as sceptics maintained that even if one was caught they would not believe it was an 'orthodox salmon' until proof on a plate evidenced 'the flavor has suffered aught by its nurture in southern waters and under a southern sky'.

Ramsbottom remained optimistic, witnessing the magnificence of 'seven distinct rises' at the same spot of his first salmon sighting, bringing back memories of boyhood days with his father and brothers in Lancashire.

On 1 April, Dr Officer took a relative, Myles Patterson, and his wife to the same area, where they saw shoals of little fish being chased by large fish. First one large fish and then a second leapt out of the water, rising at least a foot above its surface, 'affording a complete view of their whole bodies.' He judged one, about 20 inches (50 centimetres) long, to weigh six or seven pounds (about three kilograms).

'The spectacle ... was so extraordinary and unexpected, that with some feeling of bewilderment I several times asked myself whether I was not dreaming...'.

Officer presented his 'testimony' to Premier Sir Richard Dry, the first Tasmanian-born citizen to be knighted, and it was quickly relayed to all the colonies, New Zealand, and Britain: 'I think it is to be fairly inferred,' Officer boldly said, 'that the number of the returned salmon is very considerable, amply sufficient to stock not only the Derwent but all the other rivers of Tasmania, as well as those of Victoria and New Zealand, and thus our great undertaking has been entirely successful.'

As Salmon Commission Chairman and Speaker of the House of Assembly, Officer dined out on his experience. No one had 'such a favourable opportunity of observing them as was enjoyed by myself,' painting a picture of 'two great fish ... proclaiming themselves to be real salmon'. Governor Browne declared 'the successful acclimatisation of salmon in Tasmanian waters is indeed a matter for most earnest congratulation.'

Congratulations continued to shower on Youl. Following his French medals, the Irish city of Galway wanted to congratulate him on the 'unprecedented success' of introducing salmon from United Kingdom waters, including Galway, to the Southern Hemisphere. Youl received a parchment address signed by the High Sheriff and 275 others. 'It was owing to your indomitable perseverance, after repeated failures for several years, and at great cost to yourself and others ... you ultimately succeeded in conferring what, we trust, may prove to be a lasting benefit to mankind'.

At a banquet at the grand Railway Hotel, Dr Thomas Moffett, a founding professor of logic and metaphysics at Queen's College, said men of ability were usually divided into the theoretical and the practical, of thought and action. Youl had 'combined the double qualities in his own character,' with a clear and reflective intellect 'to conceive a great and original process' and an 'energetic and decisive will to consummate it'. The quest was 'almost unique in its combination of scientific knowledge and practical skill'.

Moffett recalled Jonathan Swift, 'our first great patriot', who stated that 'the man who made two blades of grass to grow where but one had grown before deserves to be considered a benefactor of his species'. Measured by this standard, Youl had indeed done good service to his country.

To hearty applause, Youl responded that having spent much of his life 'in the wilds of Australia' he could only respond 'in the plainest words possible'. Basking in the moment, he recalled that when he moved his family to London to educate his children, he reunited with 'that beautiful fish he had so often admired on the marble slabs of that great Babylon,

London'. He felt if he could introduce salmon 'he might do more good than if he remained [in Tasmania]'.

The Galway recognition was a pleasure he would remember, he said, 'to the last day of his life'. It prompted others to say 'the old saying that a Prophet has no home in his own country is applicable to the case of Mr James Youl'. The Tasmanian *Times* said he had received numerous medals and testimonials from leading continental societies but nothing from the country he benefitted. *Hear hear*, Youl might have thought.

In Hobart, hearts rose with the capture of any large fish, only to sink when they were deemed impressive specimens of trout. The *Australasian* lamented 'there is not a word said about salmon. Everybody is ... anxious to hear something more'.

No one was more anxious than Ramsbottom. When a young man reported seeing 'two large fish' in the shallows at Dunrobin Bridge, a convict probation site where the Ouse entered the Derwent 40 miles (64 kilometres) upstream, he galloped off into heavy rain with no thought but to 'capture a fish'. He saw 'unmistakable indications of the presence of salmon' and reported the spot 'as well suited for spawning as any in the world', but the river was too swollen to find any redd. After questioning the witness, he was in 'no doubt' pair of grilse had spawned. But he returned home with nothing to show for his troubles except for 'wet, exposure and fatigue'.

Melbourne's *Argus*, widely reprinted in Britain, tired of the elusive salmon. Under an editor less enthusiastic than founder Edward Wilson, the paper saw reported sightings on the Derwent as 'very vague evidence' relying on 'recollections extending back to days of boyhood ... it would be hard, indeed, for a person who has not, for mayhap 40 years, seen either a big healthy brown trout or a very young salmon ... to distinguish between the leaps, or to say from the curl in the water whether it was made by one fish or another.'

The Hobart *Mercury* snorted: 'We may have seen what we thought salmon, but have we caught them? No, and for this very good reason, that we have not tried.'

Allport had great faith in Ramsbottom – 'He would rather drown himself than tell a lie' – and Officer told the House of Assembly he could not fathom any remaining 'incredulity'. The number of sightings and authentic accounts of intelligent witnesses made it 'quite unaccountable'. He was as sure they were in the Derwent as 'I am on the floor of this House'. He had seen them with his own eyes 'as plainly as he now saw the table', and he would be a fool if he could not distinguish between them and other fish. For anyone who had seen the fish leaping even once 'the idea of mistaking a trout for a salmon appears absurd to one acquainted with both fish'.

The Salmon Commission's annual report declared 'evidence of the return of the salmon ... is complctc and irrcfragablc and must soon be confirmed by their actual capture'.

In the face of the world, Officer said he would stake his character on it. Ramsbottom would capture a salmon 'and so leave not a single incredulous person'.

19

A SALMON TRAGEDY

Despite a worsening chest complaint, Ramsbottom struggled on. In mid-November 1867, on the hottest day of the year (recorded at over 90°F (32°C) in the shade), he took 248 trout fry to Melbourne. He urged that they remain in a Royal Park pond until mid-winter before being turned into Watts Creek in the upper Yarra. For his role in 'the famous salmon ova and trout experiment', he was honoured at a Royal Zoological and Acclimatisation Society of Victoria dinner. President Dr Thomas Black lauded his 'achievement which is believed to be without precedent in the art of pisciculture'.

Ramsbottom also took more than 400 young brown trout on two steamer journeys and a three-mile overland trek to North West River Bay outside Hobart. He prepared 1200 ova for Launceston – his delivery the previous year produced only a small number of trout fry at Strathmore and they were swept away by an overflow – but wasn't well enough to make the journey himself, and apologised to the Victorians for his tardiness in delivering a report, he said: 'I have been so unwell that I could not attend to my duties as I should have wished.'

Since his desperate 80-mile (128-kilometre) return ride in pouring rain in a bid to see and capture spawning salmon at Dunrobin, and years of freezing nights and 'cold pitiless pelting rain' midwifing baby salmon and trout, he was suffering. Newspapers reported a 'violent cold' meant he was unable to personally keep watching for salmon, but illness had not diminished his conviction. 'He states that he has no doubt whatever that these were really salmon.'

But Ramsbottom was frustrated by his declining health and inability to answer the big question: why couldn't the presence of salmon be definitively settled by the capture of one?

It was the same in London. 'Have they caught a salmon yet?' was constantly put to Youl. Sensing that perhaps too few knew what a salmon looked like, he sent the Royal Society illustrations of salmon and trout. Others called for a temporary lifting of fishing restrictions designed to protect the salmon, and a reward for the first capture.

Officer assured Youl he was as certain of salmon 'as that he has a head on his shoulders' but capture was 'more easily said than done'. If the salmon had returned to their Plenty nursery they would have been more easily seen and caught, but the Derwent rendered netting unsuitable. The Commissioners nevertheless believed 'the whole experiment is beyond doubt'. And Officer reported that from the five pairs of 'salmon trout' which Ramsbottom had retained in a small pond as an experiment – 'laughed at by many people, as such a thing as had never been known to be done in England' – he now had about 120, six to eight inches long.

'There is no trusting old-world rules at the Antipodes', said Henry Francis, a Cambridge-educated judge in New South Wales and keen fly-fisher. 'Tons of this fish are annual [sic] sold in the London market as true salmon.' *Field* was astonished: 'We send a thousand ova out ... like Whittington's cat, on a venture, and our ventures comes back to us repaid a thousandfold for we can now stock any or all of our lakes with sea trout if we like, and we certainly shall like.'

Despite his poor health, Ramsbottom, along with hundreds of people, waited most of the day at the Salmon Ponds to greet Prince Alfred, but pouring rain forced an abandonment, so no royal fish or honours were dished out. Then when the prince arrived soon after in Sydney, he was the victim of the first attempted political assassination in Australia. Life was also becoming hellish for Ramsbottom. As the 1868 winter approached, friends were told he was 'suffering from a pulmonary disease and is not expected to recover ... a melancholy wreck'.

On the cusp of triumph, the *Examiner* said, readers would share its 'feeling of deep regret that Mr Ramsbottom, the intelligent and pain-staking superintendent ... is in a very precarious state of health, brought on by exposure in the energetic discharge of his duties.'

He was only 35, and after five perilous voyages across the globe and long separations – in the first seven years of marriage his salmon pursuit kept him away from Elizabeth more than half that time – he had four young children, all under eight, and a fifth on the way.

A local doctor tried spraying his throat with sulphur, an ancient remedy for consumption, with 'very questionable success'. Doing his best to labour on, Ramsbottom prepared a large can of trout ova for two esquires to carefully balance on a rugged 24-hour trek to Bothwell, from where 71 trout 'lively as kittens' were liberated in Lake Crescent and the Clyde River.

He also told the *Mercury* of eight pairs of trout spawning in the Plenty, and how he netted three to secure their spawn. Weighing nine pounds and measuring 26 inches (four kilograms and 66 centimetres), he described them as 'astonishing'.

This was his last public word on the work which had consumed him. He was granted leave of absence to travel to Sydney to escape another Tasmanian winter in the hope he might recover from his 'shattered health'. Ramsbottom farewelled Elizabeth and his children on 1 July. Seven weeks later, on 22 August 1868, he died alone and was buried in a newly opened cemetery at Rookwood, in Sydney's west. There was no public interest in Sydney, no headstone.

Tasmanian newspapers judged that 'complete devotion' had been a fatal obsession. Others felt authorities 'killed poor Ramsbottom by locating him in such a low situation, with his consumptive constitution. These irrigated lands are lovely to look at ... but never live on them unless you wish to suffer from ague (chills), asthma and the rheumatics.'

There was universal sadness Ramsbottom passed before he could reap the full honour which was his due. On the day he died the Salmon Commission had boatmen on another day and night vigil to net 'ye first salmon of ye season' and anglers were looking forward to the Derwent being thrown open in a bid to 'see our first salmon hooked'. Trout were thriving in many of the 98 Tasmanian rivers deemed suitable, and the first cooked trout would soon be served to departing Governor Browne at a Hobart Town Club dinner. Acclimatisation Society curators had also collected 3000 more trout ova for Otago, Southland and Canterbury.

The *Examiner* said that, to his last days, Ramsbottom continued to speak on salmon 'with the utmost confidence ... it was this conviction that cheered him when he felt that his days were numbered'. His name would be 'associated with one of the most remarkable of modern scientific achievements'. Transporting fish ova across the globe to hatch and become 'patriarchs of a countless posterity' might seem a small matter to some, but it was 'as complete and astounding as any of the marvellous changes wrought during the vast periods of geologic time'.

The death of the highest authority on salmon in the Southern Hemisphere was seen as 'a grievous loss to Australian pisciculture'. In London, *Field* said this loss was 'very great and difficult to make good', and his death was marked in France by the Societe Imperiale Zoologique d'Acclimatation.

Youl might have reflected that Ramsbottom and his own father had both sacrificed their health to

William Ramsbottom plaque, Salmon Ponds.

change the nature of things, each heeding the biblical call to 'rule over the fish in the sea' and 'fish for Christ'. And noted that while authorities couldn't bring themselves to pay his mother a widow's gratuity, Hobart was now more empathetic toward a pregnant widow.

Elizabeth was still in her early 30s with four children depicted as 'strangers in a strange land, and with bleeding hearts ... not well provided for', with another baby on the way. The Salmon Commission wished Ramsbottom's salary to go to Elizabeth until they found another superintendent from Britain, and parliament was asked to consider an 'act of grace' for a family 'utterly destitute' and 16,000 miles from family and friends. Colonial Secretary Richard Dry argued Ramsbottom had died doing his duty, and his small salary precluded adequate provision for his family, so he proposed £250 (£24,000 today, A$50,000). His predecessor, James Whyte, regretted the 'smallness', but hoped the colony would be in a better position to do more in a few years when the salmon success was complete.

On 9 February 1869, Elizabeth Ramsbottom left on *Windward* to return to England with Jane (nine), Robert (four), Henry (three), and Elizabeth (one). But not with five-week-old Annie, as on the morning of departure from the Salmon Ponds she died of whooping cough.

Elizabeth sailed back across the world to pick up the pieces in Clitheroe, close to her own family. She used some of the £600 ultimately provided to her to purchase four stone cottages for rental income and returned to work in a cotton factory. Soon after her eldest son Robert, named after his grandfather, died at the age of six. Surviving son Henry went on to become a pioneering teacher in Canada.

Robert Ramsbottom Sr marvelled at what his son had achieved. He recalled his son's boyhood capture of salmon on the Hodder, flowing into the Ribble near Clitheroe, at 'one of the best salmon pools in the world, called the Froth Pot', and urged Tasmanians that 'when you kill your first salmon you must christen the pool 'froth pot'.'

Ramsbottom's trout work had also inspired and captivated New Zealand. The Canterbury Acclimatisation Society put some brown trout ova from the Plenty Ponds into hatching boxes on some scrubby land it thought an ideal site for botanic gardens, ponds and lakes – the site of the now-famous Hagley Park precinct of Christchurch – where a single trout hatched in October 1867, followed by another two a few days later. The miracle births captured the country, more so when two escaped and a third was washed away in a February 1868 flood. All were presumed lost for good, but miraculously a male and female were recaptured. The Otago Society sought another 800 trout ova from Tasmania the following year, as did the Nelson and Southland Societies.

The angling excellence of New Zealand was getting its finhold. Enterprising men like curator George Clifford used a wheeled apparatus with a can of ova suspended by elastic springs from a beam and steadied by hand, adding fresh water every 15 to 30 minutes, to distribute in areas up to 80 miles (130 kilometres) from Dunedin.

But it wasn't salmon. From Dunedin, the 'Edinburgh of the south', the Otago Society, with a £1000 (£96,000 today, A$201,000) budget, sought help for a salmon shipment from Dr William Lauder Lindsay, a botanist and physician for the insane in Perth, Scotland; Scottish author-angler Thomas Tod Stoddart; and William Young, a Dunedin businessman visiting England. Lindsay, who had travelled extensively in New Zealand, quickly set about learning what he could from Youl and Frank Buckland in London, and Robert Buist at Stormontfield.

Youl said he had 'personally superintended both the successful shipments (to Tasmania) as well as the failures' and would be glad to do it for Otago. He advocated a direct voyage to Dunedin, 1000 miles (1600 kilometres) more distant than Hobart but with a latitude a few degrees further south.

He re-engaged Thomas Johnson to help build a new 55-ton (50-tonne) icehouse for the 840-ton (760-tonne) Shaw Savill clipper *Celestial Queen*. With his experience from shipments to Tasmania, an icehouse bigger than the *Norfolk* and *Lincolnshire* arrangements, Norwegian ice 'the best I have ever used', and 220,000 salmon ova, 'twice as many eggs as any one I have ever made before', Youl envisaged a significant chapter in the 'salmonisation' of the Antipodes.

He packed salmon ova from the Severn, Tyne and Tweed and ice into 315 boxes, along with 15 boxes of other ova: 9000 lake char from the King of Bavaria, 4000 sea trout from the Hodder and 1500 brown trout which Frank Buckland collected on the Gade, along with some gudgeon, carp, tench, and English oysters perhaps 'of a better class'.

'One day', Youl said, 'it was so cold and freezing ... so cold that I was glad to get into the icehouse for a warm, my hands suffering severely from freezing water and moss.' There was also friction about where any Kiwi credit might lay. Young told *Land and Water* that without Youl's advice and efforts he would not have taken on the heavy pecuniary risk involved in the undertaking. 'To him, and him only, the success of the experiment as regards the safe arrival of living ova ... must be ascribed.' He did not doubt the Otago Government would 'award him the merit which he has justly earned'.

This didn't sit well with Buckland. As a newly appointed HM Inspector of Salmon Fisheries, he wanted to be seen as a key player in a 'great and important experiment'. On the day *Celestial Queen* embarked from St Katherine's Dock, he wrote a lengthy letter to the *Times*. In the manner of a gentleman, he recognised Youl as a 'painstaking' man whose efforts gave Tasmania and New Zealand 'ample reason for gratitude', but, he said, it was he who had been pre-eminent 'with the shipment of salmon ova to New Zealand'. His version, recounted in an 1886 biography written by his brother-in-law, was that he had packed 'the first salmon and trout ova shipped to New Zealand'. He also wanted the Otago Government to recognise that his boards of river conservators enabled the ova to be secured. He gifted Otago illustrations of salmon and handmade plaster casts of salmon and trout, to be the foundation of a replica of his Museum of Economic Fish Culture in South Kensington.

When Buckland's account was reprinted in Hobart, a Youl supporter, 'Alpha', said it would 'only be doing justice' for the public to also see Young's *Land and Water* admiration of Youl as the key player.

Robert Dawbin, a Somerset farmer's son keen to find some fortune in the Antipodes, was chosen to accompany the *Celestial Queen* shipment. Youl may not have been aware his wife was Annie Baxter, who had known Youl at Symmons Plains and diarised her distaste for 'purse pride'.

Youl issued Dawbin with a step-by-step guide to handling ova, moss, ice and water. 'I trust you will implicitly carry out these instructions, and not on any account follow any suggestions you may have received from professed pisciculturists in England, or any you may receive in Otago opposed to them.' And as to Buckland's brown trout and the 'German trout ova' he strongly urged that 'on no account let them be put into the ponds with the salmon ... it would be like putting wolves into a flock of sheep'.

On a prolonged 107-day voyage, every one of Buckland's contributed fish and all but two oysters perished. When Dawbin took his lantern into the icehouse on arrival in May 1868, much of the ice had melted and boxes were in 'a state of confusion', but on unscrewing four boxes, he reported all 'very satisfactory'. Men quickly moved the remaining boxes of salmon, sea trout, brown trout and lake char for the Otago Acclimatisation Society and an experiment at Water of Leith near Dunedin.

Youl was disappointed with reports of the voyage, but 'as no less than 200,000 ova were shipped let us hope that enough young fry may be hatched to eventually be the means of stocking all the beautiful rivers of New Zealand'. But only 8000 survived the journey and barely 500 ova hatched and were deposited in the Waiwere River and pond near Leith Stream. Youl blamed the 'complete failure' on the long voyage, surviving ova being wrongly placed in ice, and floods pushing fatal fine mud into the hatcheries.. A model of Stormontfield had been sent out at his suggestion three years previously but the provincial imitation was seen as inadequate.

The Otago province nevertheless awarded Youl a large silver cup to recognise his 'very able and very valuable' service in transmitting the first salmon ova. At the same time a prominent German zoologist, Professor Carl von Siebold, lauded Youl's 'perseverance and prudence' at the Royal Bavarian Academy of Sciences. He also presciently considered that the change of climate, food and other circumstances would in time produce Antipodean salmon of 'considerable variations in shape and colouring etc'.

New Zealanders quickly resolved there should be another salmon attempt. The next year Youl packed more than 100,000 salmon ova in heavy rain onto the 719-ton *Mindora*. Buckland again annoyed Youl with a late delivery of 5000 sea trout ova, and again Youl's 'first impression was to refuse to put them in the icehouse'. But as he was told sea trout were 'beautiful fish' affording great sport, and unlike brown trout were 'not enemies but friends to the salmon', he felt it 'churlish' to refuse, although 'I have still some doubts that I acted wrongly'.

With the consignment loaded, Youl and his assistants were not slow to accept the *Mindora* captain's invite to his cabin, happy to 'take a little something' against the cold, and 'wish … our little friends in the icehouse a merry Christmas and safe and pleasant voyage to the Antipodes'.

20

DEGREES OF CIVILISATION

After a seven-year pursuit of a 'Thames-bred' salmon, Frank Buckland excitedly rushed to Gravesend to purchase the catch of what he declared a true salmon of 23 pounds. It had perhaps come across the Channel from the Rhine, or from nearby rivers, but he declared 'it is a good omen for the future salmonisation of this noble river'. The *Times* said its heart was with Buckland, but he couldn't explain the fate of millions of ova and fry deposited in Old Father Thames. 'One swallow does not make a summer,' the paper said, 'one salmon caught at Gravesend will not turn the Thames into a fine salmon river.'

And salmon sightings in the Derwent did not make it a fine salmon river. The 'have they caught a salmon yet?' chorus, Youl bemoaned, was 'constantly put to me by my friends at home and abroad, with much more blunt variations'. The *Cornwall Chronicle* said it 'is the first question put by Englishmen ... on the arrival of every mail from Australia'.

Sightings were aplenty. The Salmon Commission said more than a dozen testimonies from people of sufficient 'credit' to make their evidence trustworthy was 'the most authentic evidence'. Chairman Dr Robert Officer, ponds estate owner Robert Read and Ramsbottom's assistant John Stannard variously cited large fish jumping out of the water; 'as great a splash as a dog would have made jumping in ... I should think they were 20 pounds each ... about three feet long, the thickness of a man's thigh.'

Visiting Victorian businessman, Matthew Seal, who had transplanted perch ova from Hobart into the lake at his hometown Ballarat and was seen as a competent fishing authority, sent a telegram to Victoria that 'grilse' were numerous near the Falls at New Norfolk. 'It is true that not a salmon has been caught, but this is not to be wondered at when we consider how comparatively few in number were the young ones that were put into the river.' Given the immense extent of river 'it would be a matter for wonder almost if any of them had been seen.'

The Derwent expanse and its tributaries gave fish access to perhaps a thousand miles (1600 kilometres) of water, across an area sparsely populated, with most inhabitants not knowing a salmon. Salmon fishing by rod or net was outlawed, and the Salmon Commission had made only a few attempts with inefficient nets. Dense tea tree shrubs on the banks, sunken trees and rocks, and strong currents easily destroyed nets. The Commission lamented 'the capture of the fish in the shape of the salmon, smolt, or parr is beset with difficulties'.

A frustrated Youl conceded 'it cannot be answered that salmon have been caught', but told the *Times* saying he was 'anxious' to explain why he concluded 'the salmon is now actually naturalised'. Any fish hatched from the *Norfolk* voyage would, if they survived, now be five years old and weigh up to 40 pounds (18 kilograms). As no one had previously seen fish in the upper Derwent beyond two pounds – native mullet was the largest and never reached even one pound – reported sightings of fish larger than any trout seen in the Plenty, 'constitutes strong presumptive evidence that they are the very salmon introduced five years ago'.

Rivers favourable to the growth of trout were also favourable to salmon, he said, and noted the feast of trout, some weighing six pounds, at Governor Browne's farewell at the Hobart Town Club. A nine-and-half pounder had also been caught and returned to the Plenty. 'Surely this augurs well for the salmon. Few rivers in this country can boast such fish.' And from the *Lincolnshire* shipment, another 6000 salmon had been liberated in February 1868 and were expected to return as grilse in January 1870.

But Youl's patience was running thin. Citing a biblical proverb, he said, 'Hope deferred maketh the heart sick. After fourteen years labour I must confess to some little heart sickness at complete success not having as yet been fully proved.' His conviction was only sustained by the 'authentic evidence' of so many reported sightings, the 'amazing' success of trout, and understanding the difficulties of catching salmon. 'Such facts ... put anything like despair out of the question.' And perhaps he subscribed to the proverb's conclusion: deferral made the heart sick, 'but when the desire cometh, it is a tree of life'.

In the pursuit of confirmation, librarian and Royal Society member, Alfred Taylor, called for a 'shoot to kill' policy. 'If you tell a person that a salmon has been seen, he receives the information with a derisive smile and the usual answer "Have they caught one?"' The public would not be convinced until one was caught but netting and angling had failed. 'Let one be shot with a rifle. This is a very unsportsmanlike method of catching a salmon I admit, but the success of the experiment must be put beyond a shadow of doubt, and if it is impossible to hook a fish let one be shot.'

When Scottish lawyer and future Premier, Adye Douglas, urged a substantial £100 reward (£9,600 today, A$20,000) for a 'dead or alive' capture, Officer agreed to a personal £20 reward (£1900 today, A$3900), saying he was as sure there were salmon 'as there were stars in heaven'.

Some felt any reward cut unnecessary if the Salmon Commission was so confident. Others believed it would never be paid: the released salmon had become victims of predators, lost their bearings in the wide expanse of the Derwent and the open seas, or found their way south to the Antarctic or perhaps north to Japan where 'Tay-like' salmon had been reported. Some saw salmon on the same proof level as that of the 'sea serpent' or the Australian bunyip of Aboriginal mythology. It was also hinted darkly by one parliamentarian that individuals in Britain were 'conspiring to swindle the colony by sending out the ova of one fish when paid to send the ova of another'.

The *Cornwall Chronicle* said it was 'heresy' to cast doubt, but when the Salmon Commission could still not produce irrefutable evidence 'that a single salmon exists in Tasmanian waters', one had to accept the assertion on trust. But 'we cannot declare ourselves absolute converts' without a capture.

'An Old Trout Fisher' who had long fished rivers in Ireland, Scotland and Wales, said: 'I do not know what Mr Youl would think of us if he were told, now we have got the salmon we do not know how to catch them!'

A mainland newspaper pleaded: 'Is there no one in Tasmania who, by fishing the river Derwent, could supply Mr Youl with a 30lb or 40lb answer to the cavils of those provoking scoffers?'

Have they caught a salmon yet? The long-awaited 'yes!' was shouted on 22 October 1869. After heavy rains, a freshnet came down the Derwent where fishermen were using a large one-inch mesh seine net from a beach near Kangaroo Bluff at Bellerive, opposite Hobart. Joseph and William Ikin hauled in an unfamiliar ten-inch fish about five inches in girth. They noted its silvery scales rubbed off at the slightest touch to reveal a parr's colouring underneath, suggesting it had not long left fresh water. Half an hour later they caught a second about nine inches long. Knowing the fish were not indigenous, they made 'a shrewd guess' and quickly took their prize to town.

At the new Tasmanian Museum in Macquarie Street, Salmon Commissioners Officer, Allport and Dr Henry Butler gathered. They carefully compared the two fish with written descriptions and woodcut illustrations by William Yarrell, in his 1836 *The History of British Fishes* and 1839 *On the Growth of Salmon in Fresh Water*, and Dr Albert Gunther's more recent fish cataloguing for the British Museum – and with a preserved smolt specimen and illustrations previously sent out by Youl, Frank Buckland, and Thomas Ashworth.

From the construction, appendages and teeth examination, the Commissioners excitedly concluded that the fish were salmon smolt, offspring of the first hatched ova on their first seaward journey. They debated whether they should make an absolute declaration, but not for long. After numerous failures, close to £10,000 spending and years of 'great anxiety lest they should never return', the commissioners 'unhesitatingly ... absolutely' pronounced they were two 'veritable salmon, the product of our own Derwent'. Allport declared he was 'prepared to pledge all he was worth that the fish were the smolt of the real salmon ... spawned in the Derwent.'

Two thousand people raced to the museum to inspect the specimens while Officer headed for Government House to deliver the news that the monarch of the stream was in town. The new Governor, Charles du Cane, hurried to see for himself, and from his self-professed 'knowledge of fish' gave a vice-regal endorsement it was 'the genuine article'.

The government immediately published a special report in its *Hobart Town Gazette*. 'CAPTURE OF THE FIRST SALMON IN THE DERWENT. We have the gratifying opportunity of announcing the first capture of the salmon in the waters of the Derwent ... congratulations on the undoubted success of a great and trying experiment were generally exchanged.'

Semaphore flags relayed the news to an Acclimatisation Society party sailing in the Derwent estuary, giving cause for 'a keg of Cascade nectar ... devoted to the gods'. Loud cheering erupted in Tasmania's Parliament when the Colonial Treasurer announced two salmon had been caught. At a parliamentary banquet the 'precious treasure ... the henceforth historical smolts' were centrepieces on the main table.

The news was telegraphed to the world, and for days nothing else was talked about. In the wake of a depressed economy, spirits soared. The *Cornwall Chronicle* declared the success of 'one of the greatest, if not THE greatest experiment in pisciculture ever made in the history of the world ... [it] will be regarded throughout the universe as one of the triumphs of modern science.'

News of the catch would 'occupy the greatest minds known to us in the present day, the Press of Europe will sound the glorious triumph to every shore, and the name of Tasmania become a household world not only throughout the empire, but throughout the world. The triumph of science is complete, and that to Tasmania belongs the honour of the great achievement. Is this not a sufficient occasion for enthusiastic jubilation? We may soon make Tasmania a land unrivalled in the world for climate, for wealth and for every luxury and convenience the heart of man can desire.'

The *Mercury* congratulated 'every Tasmanian' on seeing conjecture and hope become reality. The 'little strangers' were the genesis of an industry which would see exports of fresh fish to every Australian colony and cured fish to India and the East. It envisaged wealthy sportsmen and their families visiting from throughout Australasia, Britain and Europe, further feeding the economy.

The papers imagined the joy of the 'brave spirits' like Youl and the late Ramsbottom, who 'submitting to the jeers and gloomy foreboding of sceptics manfully persevered with the ... much ridiculed ... experiment until its final success.' When the salmon and trout fisheries of Tasmania became 'one of the great facts of the Australian story ... men will be prone to ask to what enterprise and to whom the colonies were indebted for this addition to their natural wealth ... proud of their inheritance of their fame.'

Mainland papers said it was a true wonder of the last half century. But even Youl would have recognised the overreach of those who equated the salmon discovery with the introduction of sheep, whose golden fleece and meat comprised perhaps the most valuable species acclimatisation outcome the world had ever seen.

Nevertheless, the sheer ambition with salmon had been what the *Argus* described as a 'wild dream', perhaps as wild to some as 'an aerial tramway to the moon'. But through Youl, the Ramsbottoms and Edward Wilson, it had been realised.

'No sensible person can doubt that the happy event now convulsing Tasmania marks a momentous period in our natural history,' the *Argus* said. 'The gain in a scientific point of view alone is immense; for apart from the astonishing fact that ova spawned in Scotland can be packed in ice flannel like the champagne of a gourmet and conveyed sixteen thousand miles by sea to hatch safely at the antipodes ... points the way to ... still greater marvels.'

But a formal crowning needed British science to officially recognise the salmon. Not, the *Mercury* said, 'to satisfy ourselves but to satisfy the scientific world at home ... for the colony's sake and for scientific sake,' to silence all doubt and set scepticism at rest, for 'strange to say there is unbelief among us still'.

The *Examiner* admitted it was 'easy enough' at first sight to be disappointed by the two specimens. The somewhat uninteresting looking specimens had none of the rich hue of a salmon's back and side, or the glistening silvery white lustre beneath. But such vibrancy was easily lost with fish roughly handled and then dead. But the paper said those who followed the Agassiz admonition to 'look at the fish! look again!' and carefully compared the Derwent catch with woodcuts, drawings, descriptions and preserved samples from England, were at one that it 'establishes beyond doubt to the scientific observer the absolute identity'.

That identity would be cemented, it was felt, by sending home one of the fish in a specimen jaw along with Morton Allport's account of the 'Brief history of the introduction of Salmo salar *and other* Salmonidae *to the waters of Tasmania*'.

The larger specimen was preserved and retained in Hobart Town as 'the lasting memorial of our first triumph ... from what we hope we may now call our great salmon stream.' The smallest was dispatched to London, where Youl could have the joy of seeing the first descendant of his consignments, before an examination at the Zoological Society, where the *Mercury* was confident it 'would attract once more to Tasmania the attention not only of the great circle of scientific men, but of all classes of English society'.

But would the specimen be accepted as proof that man might harvest suitable waters as he had the land? That man, could turn the natural world upside down? That salmon could live in the Southern Hemisphere for the first time?

While waiting for an official verdict, Youl received a memorable letter in 1869 from the Societe Imperiale Zoologique d'Acclimation, founded in the same year he began his salmon quest 15 years before. 'Monsieur James Youl', it advised, was to be awarded 'une Grande Medaille d'Or' for his introduction of salmon in Tasmania.

Quelle joie! The greatest scientific country in the world awarding him a gold medal! In the grand city hall in Paris, society secretary Dr Jean Leon Soubeiran said, 'rarely does fiction offer more that is interesting than this true history'.

While Youl had received a silver medal in 1866 the imperial society was 'particularly struck' that he deserved high honour alongside eminent scientists for 'one of the most remarkable and laborious operations of pisciculture'. The French admired how, despite numerous failures, he 'was not the man to stop short at these reverses', exhibiting 'patience, inaccessible to despair'. And with learnings from France, the nation could also claim its part.

The Tasmanian *Times* again pointed out that Youl had still not been recognised or rewarded by his own government. His rejection of those who insisted the transporting of only fry or grown fish would succeed had saved the colony thousands of pounds. Outside a discerning few, 'no credit has even been accorded to him ... we cannot think that Mr Youl has been treated rightly.' When attempts such as *Beautiful Star* failed, 'the whole blame' was thrown on Youl. When *Norfolk* was successful, credit was thrown to others but Youl 'received no sort of recognition from anybody ... no recognition has even been rendered by this colony for his long and arduous services on their behalf.'

But indisputable recognition of true salmon was hanging on a small dead fish on its way to London. Henry Francis, a New South Wales District Court judge, and respected angler and observer of pisciculture, had not 'the shadow of doubt' about ultimate success but 'I must plead guilty to cautious reticence' given uncertain data. More practically, a New Norfolk man set a long net across the Derwent to 'forever settle this vexed question' of returning 'scaly strangers from the sea, or otherwise', but his data comprised just three local fish, a dead duck, dead platypus and an old hat.

Others questioned, if the salmon identity was not in dispute and the Ikins catch had been deemed as enough proof to win him Robert Officer's personal £20 reward, why the governor in November 1869 had approved a £30 reward for the 'detection and apprehension of the first Tasmanian [adult] salmon or grilse, dead or alive … caught in the waters of the Derwent or its tributaries'.

William Weaver, a Hobart chemist and self-professed Walton disciple born on the banks of the Wye, had no doubt the larger Ikins fish on exhibit was a salmon smolt, but suggested seeking the judgment of Francis Francis or Frank Buckland. But the suggestion was 'impertinence' to Allport, when the Salmon Commissioners were the proper judges of what should be done with historical smolts. He said Francis 'never pretended to be a naturalist and would just refer to another person'; while Buckland would give an opinion 'but I much doubt whether scientific men would rely upon it'.

After Allport despatched the smaller of the two specimens, 'this historic smolt', for Zoological Society judgment, his Royal Society's year-end celebration saluted the 'first incontestable proof of the success of the current experiment'. While some thought the specimen might yet be deemed a 'salmon trout' rather than salmon, Allport rated the chances 'about 10,000 to one against it' as some 9000 salmon had been released against only 300 'salmon trout'. 'In any event, the success of the experiment is none the less proved, as if one migratory species can succeed with a few hundred turned out, how much more likely is the other to succeed when thousands are liberated?'

There was additional cheer when Joseph Ikin produced a larger fish from near where he had netted the two smaller specimens. Again, a crowd rushed to the museum, where the *Mercury* said its true character left 'no doubt whatever', the Tasmanian *Times* lauded 'an unmistakable salmon' and the Launceston *Examiner* saw 'confirmation, if one was needed'. Even *Punch*, which had long had its share of fun at Youl's expense, declared in its year-end alphabet: 'Y is for Mr Youl, who, without any gammon/ Has stuck to his word and acclimatised salmon'.

Judgment on the small specimen, carrying Allport's 10,000-to-one odds, was on the agenda of the Zoological Society's January meeting in Hanover Square, with ornithologist-artist John Gould in the chair. He well knew Tasmania from when he and his illustrator wife Elizabeth gathered material for their seven-volume *Birds of Australia*, and corresponded with Allport. Gould had been pleased for 'our indefatigable friend Mr Youl' to hear from his geologist son Charles about dining out on trout at a farewell dinner for the governor, and his hopes to soon 'see our first salmon hooked'. But Gould Sr could not swallow salmon. He shared the trepidation about trout as 'the greatest enemies the salmon can have' and felt 'a great mistake has been made in Tasmania'. He told Frank Buckland 'I still adhere to the opinion ... that, in my time at least, we shall never hear of this partially sea-loving fish being established in the Australian waters'.

But Gould and the society's secretary, Dr Philip Scalter, also an ornithologist, resolved the specimen had to go to the British Museum of Natural History and the one person seen to be able to 'silence all doubt' and settle the question once and for all.

Dr Albert Charles Lewis Gotthilf Gunther was, effectively, the registrar of fish. The thirty-nine-year-old German had pre-eminent status in the new sciences of ichthyology and pisciculture through his cataloguing of the museum's massive fish collection and having studied and collected more fish than any man.

Gunther admitted the *Salmonidae* species was the most challenging: 'There is no other group of fishes which offers so many difficulties to the ichthyologist with regard to the distinction of the species as well as certain points on their life history as this genus.' Salmon knowledge was 'one of the most unsatisfactory portions of ichthyology' with 'almost infinite' variations dependent on age, sex, sexual development, food and the properties of their water.

In what he called a 'vast labyrinth of variation', one needed long study and a large collection for constant comparison – but he now had to pass judgment on a single nine-inch specimen from the other, unfamiliar side of the world. Identities and reputations were in his hands.

John Gould: 'A great mistake has been made'.
Photograph ©Wellcome Collection.

Dr Albert Gunther. Alamy Stock Photo.

With Youl and many others awaited Gunther's judgment, advertisements in Hobart touted 'Acclimatisation/ Salmon in Tasmania/ An Established Fact', with 'cheap fishing suits for gentlemen, hats to correspond'. Taverns and coach and boat operators anticipated the rewards of what the *Mercury* described as 'the great idea … of making Tasmania the sporting ground and touring ground of the Australias … a sort of Scotland, a favourite resort of sportsmen.'

But before the scientific blessing of another indelible tie to the Mother Country, Youl and the colonies were confronted with the unthinkable: the prospect of a break-up of the British Empire. The tyranny of distance and disinterest had long endured, but colonists were unprepared for a new Colonial Secretary, Lord Granville, withdrawing troops from New Zealand in the midst of unresolved Māori-settler wars. It was seen as a new doctrine inferring that rebellion and massacre in any of the Queen's dominions was of no concern, regardless of their British hearts and souls.

Youl could not countenance colonies like Tasmania or New Zealand relinquishing the title of 'Englishmen'. His Colonies' Association had faded, but he became vice-president of the first major 'united empire' pressure group, the Colonial Society[1], which believed the government was tending towards 'the dismemberment of the British empire'.

After chairing landmark meetings of influential colonists, he protested to the *Times* that if Britain recognised no responsibility or obligation for the colonies, even when in danger and pressing need, it pointed to an 'unfriendly, disastrous' severance. He invited all Australian colonies, New Zealand, Canada, Cape of Good Hope, Natal and Mauritius to send representatives to a conference in London.

The *Times* regarded Youl's circular as 'an epoch' but poo-poohed those 'taking fright' at the Imperial Government abandoning pretensions. With Englishmen multiplying beyond the seas by millions, it was time to end the whole 'Mother Country and Her children' notion. It was better for Englishmen and Australians if the independence of the latter had a name, just as the Dominion of Canada declared its effective independence and 'manhood'. The *Spectator*, however, supported Youl and colonists claiming a voice and the right to choose to become independent states, or territories of the American Republic, or subordinate allies of Great Britain.

Empire defenders, public disquiet, and Benjamin Disraeli's charge that the government was dismantling the empire for short-term financial gain, forced Granville to meet Youl and a deputation of former colonial secretaries, governors, and supportive MPs and aristocrats. Granville bluntly described Youl's letter as 'an attitude of antagonism to Her Majesty's Government'. As he saw it, the success of his advice to colonial governors to ignore Youl's conference call could not be 'a greater proof of the satisfactory relations existing between the colonies and myself'.

He doubted whether attempts to define the relationship more strictly, would have a strengthening effect: 'Many a man and wife, notwithstanding occasional differences, live happily together who could not do so if they called in a lawyer to define how much each was to yield on every occasion and what the terms of a possible separation should be.' So, with 'regret' he was unable to take any fresh initiative in a plan 'which

---

20. 1   The Colonial Society in London subsequently became the Royal Empire Society, then the Royal Commonwealth Society, now with 10,000 members in 100 countries).

seems already to have collapsed', thanked everyone for their 'courteous attention', and showed them the door.

The *Times* congratulated Granville for putting the colonists in their place but acknowledged Youl and 'this junta [had] at least managed to make a stir', while the *Pall Mall Gazette* saw the ostentatious indifference to colonies remaining in the empire as 'the triumph of a man who does not ... fully appreciate the questions with which he is dealing'.

Finally, Granville felt sufficiently satisfied he had made his point that the colonies needed to evince a 'manly spirit of self-reliance without which colonial self-government is a mockery', and acknowledged a duty to protect colonies until they could protect themselves. In the case of an attack, Reuters telegrammed, England would expend 'her last man, her last ship, and her last shilling in their defence'.

Not entirely convinced, Youl and others formed a new National Colonial League to press for all loyal British colonists to receive the same defence 'protection' as British subjects, resist separation, and oppose unfavourable policies. And combat the view that colonists only took 'the loaves and fishes' of the imperial system as 'suckers and not feeders' when they provided considerable political and economic value. Youl believed Australians wanted to make an Australian nation, 'but in community with Empire' with loyalty to 'home', not an independent federal republic. Youl also became an early member of a National Colonial and Emigration Society, with calls by historian-editor James Froude for the 'redundant population' of Britain to be supported to emigrate to 'loyal' Australia rather than the republic of the United States.

From the unexpected battle over colonial identity, Youl now faced the matter of salmon identity. The examination of the 'historic' specimen, sent to the British Museum, was complete.

21

A MATTER OF IDENTITY

Rugged up against the February frost, Youl was pleased to get inside the British Museum. Into the Greek revival-style building, he passed the Pantheon-inspired domed reading room at the museum's heart and headed for the office of Dr Albert Gunther.

He shook hands with the tall, wiry man with a prim bearing. Like Youl, his parents had wanted him to be a minister, but he achieved fame for his eight-volume *Catalogue of the Fishes in the British Museum*. He was known for forensic precision in identifying and naming fish, carefully measuring the numbers, sizes and proportion of scales, fins, rays, teeth, girth, proportions, vertebrae, colouring, bars, spots, eyes, and all internal features.

On his table this day was a preservation jar containing a small, discoloured, brownish fish. A small fish but with big consequences. Would the specimen be anointed by the world's greatest living fish authority as a salmon hatched in the Southern Hemisphere for the first time?

Youl held his breath. Through his wire rim glasses, Gunther looked directly at him: 'This is not a *Salmo salar*, but a trout'.

Youl was shocked, devastated and flummoxed. This was the last thing he expected. How could the scientist come to such a conclusion? The fish, Youl told him, had been taken out of water 'as briny almost as the ocean itself, and I thought trout would not live in such very salt water'.

Gunther, an occasional angler, replied that trout could sometimes be found in salt water. Youl pressed him to 'prove to me his opinion'. Taking the fish out of the jar, Gunther pointed to the scales, saying the number did not match that of a 'true salmon'. On this measure alone he was 'not in the least doubt' it was not a *Salmo salar*.

Were there any other means of confirming this opinion? 'Oh! Yes, the teeth,' Gunther replied. He then cut the string tying the jaws to examine the teeth pattern. This led to another shock for Youl, as Gunther now pronounced the specimen was not a trout but a sea trout or 'salmon trout'.

Youl was in disbelief. The world's foremost fish scientist had been asked to examine a specimen carefully sent from the other side of the world, together with a detailed account of the salmon quest history and the testimonies of numerous experienced anglers and local scientific men. Yet he had seemingly made up his mind it was a trout without a complete examination, then after a second look declared it 'salmon trout' (sea trout).

After 15 years of toil and trial, Youl could not help himself. 'It is impossible!' he protested, as the only salmon trout egg consignment ever sent to Hobart Town was in January 1866. This preserved specimen was caught in October 1869 so if it was a salmon trout it had to be about three and half years old. Gunther examined the fish again and judged it 'a miserable wretch of a fish of that age, and it would never have any power of reproducing itself'.

Youl said if this was a salmon trout then it had about 20 brothers and sisters, spawned and hatched in the ponds at Plenty, now 'beautifully grown and very handsome'. Impossible, Gunther countered. Salmon trout could not have proper ova in them without going to sea like the *Salmo salar*, and so they must be hybrids if they reproduced. 'It is easy, in a hurry, to mistake a brown trout for a sea trout, and so impregnate one with the other.'

Youl protested he had personally packed the ova, collected by Robert Ramsbottom, who knew the difference between a full-grown brown trout and a sea trout 'as well as you'.

'I put it to him categorically. Are you sure it is not a *Salmo salar*?' Gunther affirmed he was. 'Are you as certain it is a salmon trout?' Gunther said he was, and then went further, declaring neither would ever become acclimatised in Australia or New Zealand.

'He showed me a book he had published wherein he expressed his theory that the salmon would never be established in the Southern Hemisphere.' They were 'essentially an Arctic species ... inhabitants of the fresh waters of the Arctic and temperate parts of the Northern Hemisphere.'

By now Youl could barely contain his frustration. He said the newly knighted Dr Robert Officer was as certain he had seen a true salmon in the Derwent 'as he had a head on his shoulders', and at least 20 other persons had likewise seen them. 'It's wonderful how mistaken enthusiasts are,' Gunther replied.

Youl asked 'were not theorists often as mistaken?' After all, in his *Catalogue of Fishes*, Gunther acknowledged 'almost infinite variations' of *Salmonidae* depending on age, sex, development, food and waters. And that the lack of constant characteristics, even with fish obtained from the same water, let alone fish from different and unseen waters, made identification difficult.

Gunther acknowledged difficulties for anyone deciphering what he called 'this labyrinth of variations' of *Salmonidae*, but Youl could see he had a firm view about salmon in the Southern Hemisphere, and of this specimen, and he was not a man easily moved.

It was not easy for a practical man to challenge the word of an eminent scientist who had studied more fish than anyone else, and used descriptions such as 'the posterior margin of the operculum, and suboperculum is subsemicircular; when we fix three points on this curved line, viz the upper end of the gill-opening, the junction of the operculum and suboperculum, and the lower end of the suboperculum, and when we connect these three points by two cords, the two straight lines are nearly equal.'

But not even scientists universally saw Gunther as an infallible font of all knowledge. Rival ichthyologist and naturalist Dr Francis Day rejected his delineation of six migratory and six non-migratory salmon species, twice as many as Sir William Jardine in 1839 and William Yarrell in 1857. He endorsed Swiss Professor Louis Agassiz's restriction of the species to salmon, sea trout and brown trout, but each with respective varieties.

The issue was that the *Salmonidae* identification task required numerous fish to study and compare, and Gunther alone had them at the British Museum, seeking and receiving hundreds of specimens from around the world. It was difficult to challenge his reputation and word but Day, who had befriended Youl, maintained anatomical differences between young *Salmonidae* of fresh or salt water were virtually impossible to detect so he anticipated ongoing evidence of 'erroneous swearing of scientists'. He and Gunther clashed to the extent that, after the German became Keeper of Zoology at the Natural History Museum, Day shocked London by selling his extensive collection of fish to the Australian Museum for £200 (£ 20,000 today, A$41,900).

And at the Zoological Society meeting which received the Tasmanian specimen, Scottish anatomist and society prosector (post-mortem species examiner), Dr James Murie, whose work was admired by Charles Darwin, urged caution in giving decisive judgments because it was so difficult when much of the life history of the *Salmonidae* was 'conjectural'. Others were dismissive of some of Gunther's identifications, believing he was too geography based in designating migratory species such as 'Salar-sea trout of Scotland'.

But there was no christening of '*Salmo salar* of Tasmania'. Gunther had written that salmon was an exclusively Northern Hemisphere fish, endorsed Dr John Gray's view that artificial breeding of fish was 'unnatural' and without 'any practical advantage', and believed acclimatisation to be of limited possibilities. He was not going to be persuaded otherwise by one specimen, or the opinions of a former sheep farmer like Youl, or an amateur naturalist like Morton Allport.

Disappointed and frustrated, Youl returned with the specimen to his library at Clapham Park, where he and a son-in-law doctor compared it with woodcut illustrations and descriptions of salmon, including the number, size and pattern of teeth, in Yarrell's *A History of British Fishes*, and *On the Growth of the Salmon in Fresh Water*, and Jardine's *Natural History and Illustrations of the British Salmonidae*. Unable to reach a firm conclusion, the pair took his Yarrell book and magnifying glass to one of London's largest fishmongers, bemusing onlookers as he studied the size, shape and teeth of three sea trout weighing between one-and-a-half and three-and-a-half pounds and some adult Atlantic salmon. He began to feel felt that 'on the whole' perhaps it was indeed a sea trout..

He then took the fish to the Fleet Street office of *Land and Water* – which Frank Buckland had established as a rival to *Field* after a falling out with its angling editor Francis Francis – to ask naturalist-editor John Lord to examine the specimen. After an inspection with a pocket magnifying glass, Lord believed it was 'a true *Salmo salar*'. When Youl pressed Lord to 'look again', Lord did not change his judgment, but said if Youl left the fish with him he would continue to examine it. Youl then took his Yarrell book and magnifying glass to one of London's largest fishmongers, bemusing onlookers as he studied the size, shape and teeth of three sea trout weighing between one-and-a-half and three-and-a-half pounds. He thought their teeth corresponded with the specimen, but he was not completely certain. When Lord advised after a longer examination he changed his mind and considered the specimen a sea trout. But he was not surprised if Allport and Youl might have another opinion on 'so difficult a point'.

With Day telling Youl it was 'so difficult' to distinguish between any young *Salmo* species, or compare a fishmonger's freshly caught mature fish with a small, preserved specimen, Youl found it hard to fully accept Gunther's judgment.

'I cannot ... I must candidly own, make up my mind that it is a salmon trout,' Youl told the Salmon Commission. 'Because if so, it must have been ... at least three years and five months old when it was caught, and therefore ought to be half as large again as it is.' Knowing all the circumstances 'I cannot believe but that he is wrong'.

It seemed 'the height of impudence in me to challenge the conclusions of such an authority as Dr Gunther, but I do. To my mind his opinion is biased by his foregone conclusion.' But Youl reluctantly told Dr Officer he wished it to be known, publicly at least, that he had no choice but to 'concur with Dr Gunther in the opinion that the specimen first sent to England was a salmon trout and nothing more'.

In a report to the Royal Society, Gunther said the general appearance of the specimen, small size of its scales, form of caudal fin, teeth arrangement, and number of pyloric appendages[1] meant 'there cannot be the least doubt that it is an example of sea trout (*Salmo trutta*).' As Youl had maintained that the only sea trout eggs were shipped on *Lincolnshire* then 'it would follow that this example is 3 ½-years-old, and consequently what may be called a stunted individual, as a fish of that age ought to have attained a larger size, and exhibit a certain development of the sexual organs, of which no trace could be discovered in the specimen sent.'

The verdict of a 'wretched' specimen of sea trout was wretched for Youl. More so when he read Gunther's advice to Tasmania: give up the salmon quest and pursue indigenous sardine as a potential source of value. Instead of a long-awaited proclamation of success, Youl, the Salmon Commission, Royal Society, government, Tasmanians, Victorians and New Zealanders were expected to abandon their salmon dream and pursue sardines.

The judgement was a deposit on what became an extended battle over the identity of salmon, testing the reputations and relationships of men with their own ambitions.

Youl was hopeful of being credited for delivering true salmon, but Gunther gave him cause for caution. On the other hand, he had been lifted by the surprising success of the 'accidental' trout. In contrast, Allport had always been optimistic about trout, and wanted the credit for ensuring its arrival, and after his early doubts about salmon had been erased, he was heavily invested in being not only Tasmania's eminence grise of natural history, but its salmon master.

---

21. 1   *finger-like elements in the intestinal system, numbering from forty to fifty-five in salmon trout and fifty to seventy in salmon.

The issue for both men was that Gunther was an established Mr So and So who, by reputation, knew more about fish identity than anyone else in the world. And he considered the salmon quest frivolous.

Allport chaffed. It was 'his' Royal Society which had spearheaded the salmon quest by taking up Governor Denison's initial interest; receiving the various reports of fellows Bidwell, Burnett and Marwedel on ice, spawn and moss; advising terms of the reward. He had advised the government on 'management of salmon spawn and salmon fry' on a voyage, was a founding Salmon Commissioner. He had overcome Youl's objections to ensure trout's arrival and success, without which the salmon project might have foundered. Now he had to overcome Gunther's objections.

While Youl was chary of openly questioning Gunther's opinion, Allport told the Zoological Society: 'I am unable to reconcile [his] assumption, that the specimen sent to England must necessarily be 3 ½ years old, with the facts.' If Gunther had received the same fish from a Scotch river, he would have pronounced it a healthy 15 or 27-month-old smolt with nothing abnormal about its condition.

'We are asked to believe that the fish ... had found its way more than 30 miles from its birthplace and been 12 months in salt water without adding one inch to its length or one ounce to its weight.'

It was a 'forced and unnatural assumption that the specimen was an abortion.' Gunther's theory that no migratory salmon could succeed in the Southern Hemisphere would receive 'a terrible shock' if it was admitted the specimen was less than 3 ½ years old, for then it could only be the offspring of fish which had been to sea, attained maturity, and returned to spawn.

In Allport's mind, scientists in London were 'all at sea' when it came to identifying young *Salmonidae* species. But he did the gentlemanly thing and publicly thanked Gunther for his consideration. (Not least because he planned to send other fish specimens to London in return for having his name scientifically utilised, which Gunther obliged with *Anogramma allporti* (a perch), *Cheilodactylus allporti* (morwong) and *Geotria allporti* (lamprey).

Others were less gentlemanly. The *Examiner* said, 'facts are awfully stubborn things, but a theorist is a stubborner'. Having written a book declaring salmon would never succeed in the Southern Hemisphere, Gunther 'cannot and will not believe his eyes when he sees a *Salmo salar* caught in Tasmania'.

One local said Gunther's position opinion to be 'absolutely worthless'. A true scientific apostle pursuing truth, not fame or wealth, would not deny facts rather than abandon his theory. The testimony of Salmon Commissioners, dozens of experienced anglers, and the positive sightings of William Ramsbottom – who was known for being 'as careful of his facts as a microscopist' – were dismissed as 'idle tales of foolish enthusiasts'.

Judge Henry Francis told the Royal Society the scientist was suffering from 'theory on the brain'. He had fished numerous rivers and lakes in the Old Country, especially in North Wales, and was in no doubt that brown trout, salmon, and sea trout had all acclimatised. As 10-fold more salmon smolts had been liberated in 1865, two years before any sea trout, 'it is out of probability that the salmon should have failed while the sea trout has succeeded'. He had seen the two specimens in Hobart and 'I never saw the distinctive characteristics of the true salmon more clear'. Such 'brilliant' results ought not to be undervalued because of a European professor 'blinded' to an untenable theory.

The *Mercury* embraced Francis' impartiality and experience against a 'theoriser and bigoted man of science [who] could not be a party to any proof that his book was an unsound one.' Rather than accept the word of a museum scientist 'poring, glass in eye, over learned books on ichthyology', colonists preferred the opinion of men 'who have plied the angle in salmon streams of the old country from early morn to dewy eve, and whose acquaintance with the lord of the river was made in his own watery territory'.

Youl was aggrieved by the public animus. Having made it known he wanted to be seen publicly as respecting Gunther's first findings, he complained to Allport that his name had been put in company with others showing Gunther little respect. Allport's defence was that few Tasmanians had even heard of Gunther, while all but a few knew of

Youl's connection with salmon, so he thought Youl's views expressed to Officer were 'so admirably calculated to counteract the damaging effect of [Gunther's] opinion' he had not hesitated to use extracts at the Royal Society, which were duly reported by the press.

Allport said if he had not done so, the public could have felt there was something to conceal, and that the Salmon Commissioners had lost all faith in the undertaking. Worse still, it was 'quite probable' that ongoing support for the ponds and the Commission would have been 'knocked on the head' for there remained 'cavillers'. Allport assured Youl he had been careful to publicly treat Gunther's opinion with 'the greatest respect' while begging him to re-examine his opinions.

Six weeks after the capture of the two smolts, one of which Gunther had examined, the Royal Society sent a larger 13-inch specimen which Allport said was 'exactly what would be called in many English and Scottish rivers a sprod [young salmon]'. Allport asked the London Zoological Society to ensure it went to 'some competent authority', hoping the society would turn to Dr Murie, who was less dogmatic, but it again turned to Gunther.

The scientist again astonished: 'This specimen presents all the characteristics of a true salmon ... all the characters by which the true salmon is distinguished from its nearest allies.'

His wording was not entirely unequivocal, but it was enough to light the salmon fire. Youl raced to telegram Officer who excitedly advised Allport. Despite his conviction that Gunther's previous sea trout judgment showed 'no man can with certainty tell one species from the other to the smolt stage', Allport now immediately reported to the *Mercury*: 'The large smolt caught in the Derwent in December 1869 ... has now been pronounced a true salmon by Dr Gunther.'

He told the Royal Society, 'Dr Gunther at once declared [it] was a true salmon ... in plain English it was a salmon'. The blunt version was telegraphed around the colonies, New Zealand, Britain, and the rest of the world.

Youl and Tasmania expected this would finally close the book on salmon identity – but this was not to be.

22

SALMON AT LAST

A continuous of 'Another Salmon' sightings, without capture, frustrated even those of the faith. As the *Australasian* reported, 'we have again the old story that the salmon is to be seen in the Derwent. It is a pity that one cannot be caught; they are worthless if they cannot be turned to account.'

The absence made Youl anxious about ultimate and complete success, but Morton Allport told him he wondered why 'you do not regard the capture of smolts as so conclusive of success as I do'. The increasingly confident Allport assured him 'I get as much chaffed as you do' but Youl should 'let them laugh ... soon the very men who are for now disbelievers will endeavour to assert they foretold success.'

Allport was 'as firmly convinced as a man can be, that thousands of smolts went to sea the previous spring', telling John Gould that 'of their return from the sea I cannot doubt'. He told Youl he had previously been cautious in accepting 'wild' statements about salmon being seen and holding out hopes which might be doomed to disappointment, 'but I am on full consideration as firmly convinced as a man can be that these smolts were from ova laid by grilse or salmon in the Derwent'.

Dr Gunther had inspected two more small specimens sent by Allport, and while he said the pyloric appendages count of one was aligned with river trout, he told Youl all the external features of both 'agree ... with salmon parr and are very different from young trout or sea trout'. For Allport, the mixed judgements meant three possible conclusions: salmon and sea trout up to the smolt stage could not be distinguished from one another by any known scientific process; sea trout detained unnaturally in fresh water had spawned and produced a true salmon; or Tasmania had accidentally and unknowingly detained true salmon in fresh water and it had thrived and bred. The first was the most convincing.

Desperate for a catch, the government decided to allow a salmon 'posse'. It re-opened the Plenty, with £1 licenses, to a select dozen anglers, including the Salmon Commissioners, Ponds estate owner Robert Read, the Pond keeper, and some respected anglers from Victoria. Anyone else found illegally fishing for salmon would be fined £3 and have their rod and tackle confiscated.

Among the first to take advantage were Dr William Whitcombe (who had hatched perch ova, supplied by Allport, in his Ballarat garden) with friend Henry Cane (a pioneering malt maker) on a November 1870 fishing holiday. When they reached the upper Derwent, Cane declared: 'By jove! A rare salmon stream, there'll be some whoppers taken out of that water.'

They didn't land any salmon whoppers, but were rewarded with five trout, the largest a two-foot 'boomer' weighing just under six pounds. They sent this to Allport to photograph, and Cane proudly wrote to his reverend father in Nottinghamshire: 'As I have become famous by being the first Australian to catch an English trout, I send you a photo of the latest I caught with the rod you sent me.'

Whitcombe took issue with some suggesting the Tasmanian specimens were 'bull trout', not brown trout. He could not accept that Youl, Buckland and the Ramsbottoms 'deliberately deceived' the government when they assured only three species of fish ova (salmon, sea trout, and trout) were ever sent. Suggestions of dishonesty or gross carelessness he said would cause Youl and the others to be 'riled' if they heard such suggestions.

Youl saw a copy of the Cane trout photograph in the window of *Field*'s office. It coincided with a two-part series, *Salmon as a Colonist*, by Arthur Nichols, in *Field*'s *Quarterly Magazine and Review*. Youl had given Nichols, a natural history writer-lecturer, access to his collection of letters, reports, and newspaper accounts. Nichols, perhaps at the instigation of Youl, said he wanted to detail the story of 'one of the most brilliant triumphs ever achieved in the establishment of a species in regions unfamiliar to it. Notwithstanding that the ocean is an open highway to the salmon, and that it is distributed over an immense area in the Northern Hemisphere, it could never have extended its range to southern latitudes without the intervention of man.'

Two months later, Bush Inn owner Joseph Oakley reported that a shoal of some 40 or more salmon, weighing from 20 to 30 pounds, were seen in the Derwent. Then word swept Hobart that a 'real live salmon' had been netted near the Plenty. The *Mercury* thought this would finally put the salmon question for ever to rest, but other locals wondered if this was just another 'rumour ... another cheap translation of the well-known fable'.

Recent captures had seemed promising, only to be judged to be well-fed and handsome brown trout or sea trout. A crowd waited anxiously for Sir Robert Officer to arrive with the latest catch, a 24-inch, seven-pound female (60 centimetres and 3 kilograms) seen to present different features, with red trout-like spots absent, its belly 'more silvery', and head 'neater'. In all, the general cut and shape was 'more like the salmon proper'. But Allport wanted to closely inspect it against the criteria of William Yarrell and Gunther, particularly the teeth and fin rays.

The crowd groaned when Allport declared it another splendid specimen of sea trout, and again when it heard he had immediately sent the fish to the dining table of Governor du Cane for 'the most practical test of its value to this colony'. The governor happily declared it 'equal in flavour to any caught in a salmon river in the United Kingdom', but the matter was denounced as 'an unparalleled act of flunkeyism'. Whether it was truly a sea trout, a contributor called Piscis wrote, 'toadyism' had deprived the public – the taxpayers funding the quest – from inspecting what may have been the first real salmon captured. In London, *Field* was aghast that after all the identity frustrations with Gunther, the Salmon Commissioners had not sent the specimen to him for definitive judgment: 'They have just let an excellent opportunity to slip – down the throat of His Excellency the Governor.' *Field* could not believe the 'disastrous mistake'. For a brief dining pleasure 'they have deprived the world of the lasting enjoyment that would have accrued from the publication of a decisive account of the fish by a competent authority'. Youl was not amused.

Officer continued to report salmon 'can be seen running up and down the Derwent', and the capture of any strange fish bearing any similarity to salmon continued to cause 'a great stir', as the *Mercury* described. But it felt it 'very strange that none of the fish so often seen are ever caught'. The paper pleaded for salmon to have its say and end the agony.

'If the salmon of the Derwent are at all cognisant of the doubts entertained as to their existence, and grateful for the public spirit which brought their progenitors to the noble waters they now inhabit, we could almost imagine that they would "wish" one of their number would throw himself upon the Meredithian altar, a sacrifice to enable doubts to be removed.

'How the sea trout could have escaped its enemies and the salmon succumbed to them, is one of those mysteries which tends to increase the hope, if not the belief, that the salmon is an inhabitant of the Derwent, the absolute incredulity of some ... notwithstanding.'

Despite the 'savants [being] divided as usual', the Hobart correspondent of the *Argus* was no longer incredulous. 'When I cry "the Salmon, the Salmon!" do not remind me of the fable of the wolf and the boy', he wrote. While he had some doubts about Officer's statement of salmon continually running in the Derwent, he was convinced a fish which leapt into a boat near Officer's home was a grilse sacrificing itself to sustain the credit of acclimatisers and help settle a parliamentary vote for funding of the Salmon Ponds. The man who found the fish came from Ayshire 'where he landed many a salmon and grilse from the pools and salmon runs of the classic Doon ... he has no doubt whatever as to the real character of the stranger'. The four-to-five-pound female evidenced an olive-coloured back and silvery belly of a fish fresh from the sea, laden with spawn, its eyes forward like a salmon, with fins and tail of a handsome salmon. And at the dinner-table test, the colour and flavour of the flesh showed it to be 'a veritable grilse'. With this and similar reports, the correspondent declared, 'I am now ... a believer in the success of the acclimatisers, and in the absolute fact of salmon being in the Derwent.'

To further aid Tasmanians in the identification parade, Allport asked Youl to secure 'true to life' casts of 'salmon and salmon trout' fish from one of the rivers where ova had originally been sourced. They would 'enable me to dispel all doubts as to our grilse and salmon when caught as they assuredly will be before long'. He would pay £10 for Frank Buckland to make the casts and Henry Rolfe, famous for his hand-coloured studies of salmon and trout, to paint them.

Allport was incensed by a *Mercury* report suggesting it was Buckland who had initiated the casts 'to enable the authorities to decide whether the fish taken are salmon or not'. Based on his numerous examinations of 'adult salmon from different rivers' and being well acquainted with the descriptions of Yarrell and Gunther, Allport fumed, 'I never doubted my own ability to identify a salmon when I saw it.' It was 'offensive' for anyone to think he needed casts to decide salmon identity. He sought them because they might be useful for those who had never seen a salmon and were ignorant of scientific descriptions, yet volunteered their opinions 'with an unblushing confidence apt to deceive an unwary public'. The *Mercury* editor sated Allport's annoyance by adding a footnote that to him 'the colonists will feel specially indebted'.

A prolific letter-writer, Allport apologised to some for pressing his salmon views but asked they 'please bear in mind that I have been ... actively engaged in pisciculture and for a much longer period studied the natural history of fish.' He was not shy about issuing character references on the 'twaddle', 'nonsense' and 'idiocy' of others. As for those who could seemingly distinguish a grilse of salmon and sea trout in glass jars, without close reference to anatomical details, they clearly knew more than Yarrell and Gunther, 'so I must hide my diminished head and decline to discuss the subject with such heaven-born geniuses'.

Gerrard Krefft, the German-born curator of the Australian Museum in Sydney, could not understand having casts for 'true to life' comparisons when an original, preserved in ice, salt or spirits, could be easily sent. He acknowledged the differentiation of young salmonids was difficult, but Gunther was 'our greatest authority'.

'The question at issue.' Krefft said, 'is, have the Tasmanians the true *Salmo salar*, or have they mistaken the great sea trout for it?' The chief differences appeared to be size, scales and taste, but 'at present we cannot accept taste as a test'.

Disputes over fish identities caught the eye of Jules Verne. In his newest exploration of technology freeing man into perilous spaces, *Twenty Thousand Leagues Under the Sea*, inspired by the fisheries exposition in Paris which Youl had attended, Verne's main character asks a fisherman if he knows about fish classifications. The man replied: 'Sure I do. They're classified into fish we eat and fish we don't eat!'

The deep Tasmanian desire for true salmon to catch, sell and eat, and for its own refreshed identity, also evident to the most famous cultural figure to visit the Antipodes. Anthony Trollope, regarded by some as 'the most popular novelist of the day', had been urged by his friend Charles Dickens to visit.

With photographs in shop windows so locals could identity him, Trollope was bombarded by locals wanting to know 'what do you think of us? What do you think of our colony?' He shrewdly replied: 'Tell me what you think of yourselves.'

Trollope saw an island 'more English than is England herself ... Tasmanians in their loyalty are almost English mad'. His other indelible impressions were the continued 'odour and flavour' of convictism, and the 'intense interest' in salmon.

As for the unnatural efforts to capture salmon, Trollope enjoyed dining out on 'better trout in Tasmania than ever I did in England ... I ate them again and again, with great satisfaction', but despite 'so-called salmon ponds' – which a Victorian visitor said ought to be more truthfully known as the trout ponds – no one had caught a grown Tasmanian salmon. After talking to Allport, however, Trollope 'found myself ready to swear ... that there must be salmon' although a great but 'adverse' scientific authority had declared it impossible.

At the same time, Youl received advice from Gunther about specimens of adult sea trout which had never gone to sea, and two young progenies. The scientist saw a larger female as agreeing 'perfectly' with the migratory sea trout, except its smaller pyloric appendage count was something he had only seen in river trout and hybrids of river and sea trout, and its colour, spots and markings were 'peculiar'. Allport again expressed 'the sense of great obligation to Dr Gunther, but rejected the conclusion. To his eye, apart from the appendage count and some colour peculiarities – which could be explained by variations appearing and disappearing, the effects of unnatural detention in fresh water, or time in a preservation jar – it perfectly matched the 'salmon trout' descriptions, 'so can anyone doubt this ... was a pure salmon trout'? Having been hatched from an egg in England, if it was a 'hybrid' then those who had secured the ova 'most wilfully and maliciously have played a trick upon all engaged at great expense', a conclusion not easily reached by anyone who knew Youl and William Ramsbottom.

The only conclusion one could draw, Allport said, was that in their early stages of growth the migratory species of salmon still 'cannot, with any degree of certainty, be distinguished from one another'. Nevertheless, Allport was confident enough in his own ability to declare more specimens to be true salmon smolts, even though they were deemed too small to be of enough classification value to send to London.

Allport complained to Youl that fishermen had admitted taking numerous smolts in seine nets for two years, but the 'rascals' eschewed claiming the £30 reward because they anticipated it would inevitably rise to £100. Judge Francis lamented 'it is a thousand pities that such infants should be caught. For aught we know, thousands of similar fish may have been wantonly destroyed.'

Allport continued to maintain all the specimens sent to Gunther in 1869 were of one species. It was clear migratory *Salmonidae* were yearly breeding in the Derwent, and perhaps some of its tributaries, and thus to him the odds heavily favoured them being salmon and not sea trout. A shoal of large fish seen upstream, he said, was probably salmon on their way home from the sea, not overgrown trout as they did not appear in a shoal, or sea fish as they were well beyond brackish water. 'In short,' declared Allport, 'they could not conceivably have been any fish but salmon.'

He told Youl that while absolute proof was not forthcoming, he was 'as confident as ever that Ramsbottom could not be mistaken when he reported seeing them and that their numbers will soon make catching them easy'.

But by the end of 1872 the *Mercury* admitted belief had 'sank to a minimum'. The inconvenient truth was that three years on the £30 reward for the apprehension of the first adult Tasmanian salmon, 'dead or alive', remained unclaimed. Salmon sightings, a Salmon Commission and the Salmon Ponds did not a salmon make.

A year on, the flood of sightings but drought of captures led the *Tasmanian* to declare that either all 'the fishermen of the Derwent are truly denizens of Sleepy Hollow, or the salmon are not "all there".'

But a week later Sleepy Hollow was well and truly roused. An Irish emigrant was walking beside the Derwent about 300 yards (270 metres) on the Hobart side of Bridgewater where the fresh waters of the upper Derwent met the brackishness of its massive estuary. Joseph Cronly (also Cronley), a stone cutter who had arrived as part of the 1850s Irish exodus, was in the area as an inspector of masonry in the construction of a new railway line from Hobart to Launceston.

Alerted by workers who had observed splashing and whirling in a shallow pool under two willows at a railway culvert, Cronly saw a large fish, stranded after the tide receded. He raced home to grab a rod and reel he normally used to catch eel and waded in between the fish and the river to prevent its escape. After the fish refused bait 'big enough to catch an alligator' Cronly used his line and naked hook in a series of 'snatches' to jerk it out.

The two-pound 14-ounce (1.3 kilogram) female was quickly sent to Hobart. Allport was excited to see a fish 'unlike any of the large fish hitherto taken', and he again turned to the minute classification criteria of Gunther and Yarrell. He first concluded the catch was not a brown trout, based on many characteristics and the fact it was found seven miles further downstream than any other trout and was in salt water. He then turned to the more difficult task of resolving which of the migratory salmon it was, salmon or 'salmon trout'.

Allport was confident about reaching a verdict on a specimen much larger than the 'minute, almost imperceptible and perhaps doubtful differences' of younger smolts seen by Gunther. He sensed 'the present capture differs most materially from any specimen yet seen in Tasmania'. And against all the minute tests this specimen, unlike any of the large fish hitherto taken 'coincides exactly with the description of his true salmon', especially its fin rays, scales and absence of vomerine teeth, 'a distinction entirely confined to true salmon'. He was perfectly satisfied it was 'a true salmon in the grilse stage … no hesitation in pronouncing it an undoubted salmon grilse.' He christened it the 'Derwent Grilse'. To him, it was a larger version of the specimen which Gunther had previously said 'has all the characteristics of a true salmon' and so every specimen was send to Gunther. 'Can there be any doubt?' he said, especially when it was remembered that 9000 salmon fry had been liberated against only 200 'salmon trout'.

Allport sensed his big moment. He wrote to Baron Ferdinand von Mueller, the government botanist in Melbourne, that the fish would 'greatly astonish those scientific men who doubted the final success of the experiment … I fear true science will never meet its proper reward in a young country.' He advised European museums and told the Zoological Society of London that 'all doubt as to ultimate and complete success has vanished at this end of the world'. And to Charles Gould, son of John Gould, that 'the scientific world [is] all round now … it is seldom man gets such a chance.'

The news saw people flock all day to the museum to see the 'Derwent Grilse', initially on a plate and then in spirits.

'A SALMON AT LAST', the *Mercury* headlined, although after years of 'hoping against hope' it remained cautious, stating: 'or at least it is really believed to be, by those who affect an acquaintance with the outlines and general appearance of the desired fish'.

While the *Tasmanian Tribune* declared that as 'an old saw says, the proof of the pudding lies in the eating of it', most saw proof enough. The Salmon Commission accepted it as salmon 'without a doubt', the government awarded Cronly the £30 reward, and the specimen was shipped to London for examination. 'Not for identification', Allport insisted, 'but as a conclusive proof of the ultimate and entire success of the greatest experiment in acclimatisation that has ever been attempted.'

He recalled that leading scientific men in England had long predicted the 'disastrous failure of the whole salmon experiment', even for trout. But dozens of trout rivers in Tasmania, Victoria and New Zealand were 'convincing sceptics that they may have been a trifle too hasty in settling Nature's laws for her'. The 'Derwent Grilse' was the crowning salmon proof.

Allport was irked by the *Mercury* casting 'doubt, if not ridicule' on his salmon determination while giving credence to doubters not qualified to have an opinion, including one who cited the 'coarseness of the [specimen] head'. 'If he will just take a turn round the nearest lunatic asylum, he will find number of both coarse and fine heads, yet they all belong to "Homines sapientes".'

The *Mercury* editor, who saw Allport as 'too sensitive' in his 'petulant sneer', made it clear he was becoming tired of the salmon saga. Those who hinted doubts were 'stigmatised as craven and contemptible detractors', while those who merely expressed a temperate belief rather than 'wild ecstasies' were taunted with ridicule. While not qualified to give a decided judgment, the *Mercury* was willing to 'admit the accuracy' of Allport's grilse judgment but made a final point: one salmon did not prove the success of species acclimatisation.

Nevertheless, *Punch* in Melbourne suggested Hobart might be renamed Salmon Ville. And that whatever the virtues of Saint Salmon it was responsible for a Dickensian vice, prevalent among husbands returning home late and incoherent, explaining 'my dear, it must have been the salmon'.

When Youl heard of the 'Derwent Grilse' capture, he wasn't fussed that his long-awaited adult salmon had been caught, not by an esquire of the Walton brotherhood craftily using a rod and line, but an eel fisherman snatching a fish to the riverbank. He just saw the Holy Grail and immediately sent a note to the *Times*. 'Will you oblige me by conveying intelligence to your numerous readers in England, Scotland, Ireland and the Continent ... that the salmon has been successfully naturalised in the rivers of Tasmania, and a government reward for the first catch had been paid. Thus, one of the greatest feats in acclimatisation has been accomplished at the Antipodes.'

Francis Francis described Youl as being 'overjoyed ... and well he may be given the time, trouble, expense and anxiety that has cost him'. While the specimen would be studied by the savants at the British Museum, Francis said he did not need their judgment. 'It is beyond dispute that the salmon in the Derwent have bred in the river... the grilse is undoubtedly a native-born Tasmanian grilse.' Probably there were 'thousands of brothers and sisters about somewhere ... I have never wavered in the belief that the fish leaping in the Derwent were actual salmon, feeling that would come to net or hook someday.'

Youl's name, he wrote in *Field*, would be remembered in Australia and New Zealand when men were 'again spearing salmon from a broken pier of London Bridge'. He called for due reward, especially after foreign governments had awarded gold medals and certifications of approbation. If he was British, he said, the government would have probably rewarded him with an annual £200 (£18,500 today, A$39,000) pension.

The news was celebrated in Britain as 'a real salmon in Tasmania at last'. It came with reports of Australians having their first victory over a touring English cricket team led by the great WG Grace. The world was truly being turned upside down.

23

TIME & TEMPERATURES RUNNING

Following the 'Derwent Grilse' capture, Youl looked forward to the April 1874 meeting of the Zoological Society of London and the reading of a paper by Morton Allport, *On the Introduction of Salmon to the Waters of Tasmania*.

Three months earlier, Youl complained to Allport about 'misstatements, mistakes and inaccuracies' in his earlier 15-page paper, *Brief history of the Introduction of Salmon and other Salmonidae to the Waters of Tasmania*, which formed part of the Salmon Commission's 1873 report to parliament.

Allport referenced Youl's conviction, since 1859, of ultimate success after having the Australian Association to take up the cause, and exhibiting 'untiring zeal and industry in the management of such portions of the various attempts as had to be conducted in Great Britain'.

But Youl was irritated that Allport did not recognise his quest had begun in 1854. Allport also said the packing of ova in wet moss in a box in the *Beautiful Star* icehouse was suggested by Wenham Lake Ice manager Henry Moscrop, when Youl maintained the 'crucial' moss method was his 'inspiration' after a visit to Paris, and he had 'often' rejected Moscrop's claim. And while he had previously dismissed the *S.Curling* attempt as a failure from which 'no practical knowledge' resulted, he was now framing his use of the icehouse as an integral step toward final success.

Waiting for any Allport reply, Youl was keen to hear a reading of his updated paper. He was pleased to hear the declaration the 'Derwent Grilse' was 'incontestable proof of the greatest experiment in acclimatisation that the world has ever seen'. Tasmanians were often accused of apathy and want of energy, Allport said, but for 20 years the colony had 'persevered with unflinching determination, in spite of adverse scientific opinions, at a cost of many thousands of pounds, through several total failures, and ending at last in thorough and triumphant success.'

But Youl was irked the audience heard nothing of his efforts. Rather, Allport insisted the role of his Royal Society of Tasmania was central and ought never be forgotten. Its published proceedings, Allport said, proved that the idea of introducing British fish had its origin among fellows of the society, which in turn had 'strongly urged' the government to make an attempt, and had advised on the attempts, and closely recorded every stage.

On the conclusion of the reading, Youl rose to his feet, and made clear his disappointment that 'due credit' had not been given to those who had been 'most active in accomplishing the acclimatisation of the fish': himself, Edward Wilson and others who found the £700 for the first salmon attempt, and Robert Ramsbottom. For the trout success, Francis Francis and Frank Buckland also deserved credit.

Allport was indignant when he received Youl's January letter about his first history paper, describing it as a 'virulent attack'. He told his son he had given Youl 'a dose in reply', and told Sir Robert Officer that 'I have a strong notion that the next time he wants to let off his own importance he will do it on someone else'. Allport told Youl his complaints 'caused me no little surprise'. He rejected the significance of packing ova in moss, as any elastic, fibrous material, insoluble in water, could have answered as well. 'You might as reasonably claim that the species of the wood used in the boxes, or the size of those boxes, as an element material to success.' The essential breakthrough, Allport said, was the realisation of Nature's law that a body above the freezing point immersed in ice could not fall below the freezing point. This notion of John Bidwell was first noted at a Royal Society meeting in 1852, and Youl also seemed unaware that the same year saw Society fellow James Burnett also suggest using ice.

And Allport was 'astonished' by Youl stating that moss packing method suggested itself to him after meeting Zepherin Gerbe in Paris in 1861, when the French had promoted that very method to Society member Ernst Marwedel on his 1857 visit. And Allport said he had referenced this advice 'in one of my excellent reports on the salmon experiment' which he had forwarded to Youl.

As for the Moscrop dispute, Allport said he could not possibly know of Youl's views when he had not publicly challenged Moscrop's version, first raised by him in the *Times* in 1863.

On the trout question, Allport prosecuted his desire for credit. He maintained he had instigated a push for trout and ensured William Ramsbottom ignored Youl's 'considerable opposition' to have trout in Tasmania and then stood firm when the Acclimatisation Society in Melbourne declared it had a letter gifting it the ova. Moreover, without the delivery and success of Youl's 'unwanted' trout, there would have been no support for the *Lincolnshire* shipment, 'the ponds would have long since been dismantled', the Salmon Commission 'broken up', and the whole salmon experiment 'imperiled [sic] because the Derwent would have been given up without a murmur to the seine net men and poachers'.

In closing, Allport said he would be 'very glad' to see a more complete account … from your pen and trust you will find it easier to please everyone'.

Youl's simmering eased in May 1874, when Queen Victoria awarded him a Companionship of the Order of St Michael and St George. *Field* said the recognition of his services in having salmon introduced into

Australia was 'acceptable to the world of science ... it is no more than was deserved.' But it noted that while he had been medalled by foreign governments and the Queen, he seemed to have 'earned nothing beyond ... detraction ... from the colony he has served.'

While Youl wondered if *Field's* words would be heeded, he was more anxious that his salmon quest might be overtaken by New Zealanders and Victorians pressing their case across the Pacific, where 'Johnny Fishseed' had been enthusiastically adopted in America.

As the American population rose from 13 million in 1830 to around 40 million by 1870, excessive and indiscriminate fishing, mining, dam-building, manufacturing, and rural and urban development led to more areas facing an 'almost fishless state' without salmon, trout and lake shad. George Marsh, a pioneer of conservation as a movement, said the country might have to accept crippled condition of the land due to 'human progress', but might 'still do something to recover at least a share of the abundance which ... the watery kingdom afforded'.

Men who first explored artificial propagation, such as Seth Green, were initially seen as 'crazy' before becoming known as fathers of American fish culture. Scepticism over fish being 'stall-fed' and fattened like animals was trumped by money. After nine years of study, Stephen Ainsworth, a New York state nursery-orchard owner, claimed a USD$47,000 (USD$1.3 million today, A$2 million) investment in a trout hatchery would, after four years, return USD$468,000 (USD$13 million today, A$20.7 million).

Following the bloody Civil War, the Federal Department of Agriculture declared artificial fish propagation of 'national importance'. President Ulysses S Grant appointed Spencer Baird, a Smithsonian Institution zoologist, as head of the first national fish commission. Baird could see salmon disappearing in the face of its biggest enemy 'as did the buffalo of the plains and the Indian of California ... Wherever the white man plants his foot, and the so-called civilisation of a country is begun, inhabitants of the air, land, and the water, begin to disappear'. Gripped between the murderous greed of fishermen and advancing civilisation there was 'no altar of refuge for the salmon in this country ... what hope is there for the salmon in the end?'

Spencer Baird: Man is salmon's worst enemy. Photograph William Bell. Wikimedia Public Domain.

The only hope lay in emulating Youl's quest by transporting fish ova across the country by whatever means necessary, using stage, rail, steamer, wagon and sled, and the efforts of men like former pastor-turned hatchery operator, Livingstone Stone, who took a 3000-mile trek (4800 kilometres) to northern California's remote McCloud River in 1872 to establish Baird Station, the first federal fish hatchery, with a focus on salmon. He was helped by the Winnemem Wintu people, who believed that on their creation 'we were helpless and unable to speak. It was salmon, the Nur, who took pity on us humans and gave us their voice. In return, we promised to always speak for them'. Stone shipped two million salmon eggs in layers of moss, in constantly refreshed milk cans of water to replenish the east coast, while millions of east coast shad, striped bass and Penobscot River salmon, and later brook and brown trout, made their way west.

Youl could see the Americans becoming empathetic to efforts by New Zealand and Victoria to win the salmon race at a time when

Tasmania's triumph remained in the balance. Experienced anglers and many a 'canny Scot' visitor continued to report 'true salmon', and Allport continued dissecting and studying shapes, colours, spots, scales, pyloric appendages, teeth or tail shapes – forked in salmon, more square in sea trout – to discern identities. Specimens of 'true salmon' continued to be identified and sent to Dr Gunther, who continued to issue mostly sea trout judgments.

Gunther did place one specimen in the British Museum's public gallery 'as evidence of the remarkable success which has attended the efforts of the colony to introduce salmonoids', leading Allport to state it was gratifying scientific men had changed their opinions on the success of at least one species of migratory salmon, as while sea trout were commercially somewhat inferior to salmon 'they are worth 50 times more to the colony than the cost of the experiment'.

The *Times* regretted the absence of an authentic, adult salmon capture meant 'we are, unfortunately, not able as yet to pronounce that the "salmonisation"... is an established fact', but Allport said Gunther had not changed 'my unshaken belief that if 200 salmon trout have succeeded, 10,000 salmon turned out under precisely similar conditions must have succeeded also'.

Youl hoped it was true, but with Tasmania not engaging in more shipments, and Baird warming to the dispatch of the first quinnat or chinook salmon ova from California to New Zealand, he was back on the Kiwi quest. At East India Docks in 1873, packing 120,000 ova on the *Oberon*, he invited Judge Henry Francis to accompany him down a hatchway to the darkness of the lower deck to see his icehouse. 'Taking the lamp from the stevedore, Mr Youl descends the ladder and bid me follow' into a chamber the size of a small dining room, with blocks of ice making for a 'slippery and treacherous landing place from the ladder'. The judge noted 'the strength of public spirit which makes Mr Youl, now a good deal past middle age, encounter the dangers of ladders, hatchways and arctic temperature, instead of sitting quietly at home to read the *Times*'.

Youl was disappointed *Oberon* landed only about 10,000 living eggs, from which only 96 hatchlings survived. But he got some joy from ova he hatched at home. The final survivor, a 'lively little fellow', was adopted by the Youl household for 14 months before it was time for it to 'graduate' at the new Brighton aquarium. Director Henry Lee said: 'I hesitated to accept the responsibility ... from his birth he has been the pet of Mr Youl's family, and, from kitchen to parlour, the household were as sorry to part with him as if he were an only child about to be sent, for the first time, to a boarding-school.'

Youl was sorrier to have Frank Buckland criticise the *Oberon* outcome and step into the New Zealand arena. The Inspector of Salmon Fisheries of England and Wales acknowledged 'no man in the world' knew better than Youl how to pack salmon eggs, but it was 'a matter of regret to all' that rather than being an official international government shipment, of which he would have been in charge, the shipment was entrusted 'to one individual ... (and) I make bold to assert that Mr Youl's experience in the actual breeding of salmon has not been very great.'

Buckland said trout had succeeded in Tasmania, 'mainly through the instrumentality of your humble servant' and 'I cannot see why salmon should not succeed in New Zealand'. He offered his services, sent detailed instructions, and presented the Canterbury Museum with a cast of a full-grown salmon which 'may prevent a repetition of the difficulty experienced in Tasmania as to the identification of the true salmon'.

Youl didn't warm to the posturing. When Buckland suffered a failed bid to transplant 300,000 salmon ova on *Timaru* in 1875, Youl told *Times* readers he 'sympathised' but it was 'a pity' Buckland had not adopted his successful packing technique. 'My packing may be compared to a nest made by a blackbird; Mr Buckland's to that made from sticks by a rook.' While Youl put a double layer of charcoal and ice at the bottom of perforated boxes before creating a nest of moss and covering the ova with more moss and ice, Buckland eschewed the charcoal-ice foundation and put alternate layers of ova and moss in larger boxes in a tub of water.

Buckland conceded the difficulty of the task, but 'I by no means acknowledge myself defeated by the salmon or its eggs'. He would not rest until he 'solved the problem'.

Youl spent the 1875 Christmas season readying for another shipment to Otago when he learned that a man with deep pockets wanted Buckland to send Atlantic salmon ova to Melbourne. Irish-born Samuel Wilson, with enormous stations holding 600,000 sheep, was perhaps the world's biggest sheep owner. He had an enthusiasm for acclimatisation and had hatched and distributed San Francisco fry by Cobb & Co. coaches to about 25 Victorian rivers. Now he sent Buckland £1000 for a British shipment.

Perhaps out of desire to ensure 'his' Atlantic salmon from Britain would not be bested by Pacific salmon from republican America, or that his packing 'credit' might be threatened, Youl agreed to a joint steamer shipment of 175,000 ova to Victoria and New Zealand on *Durham*, but only on the condition he and Buckland would pack their own ova. He described it as 'a trial for the benefit of the Australians as to whether the packing by Mr Buckland or myself is the best'.

Publicly, he joshed that he hoped Buckland's would turn out the best as 'in that case I shall not be called upon to pack any more for at my age it is high time I gave up the risk of standing about the miserably cold docks superintending every detail so necessary to success. I know from painful

Samuel Wilson, c. 1861. Studio portrait by Batchelder & O'Neill (Melbourne, Victoria). Courtesy Parliament of Victoria.

experience how difficult it is to get salmon ova after Christmas and the packing of them in January with from 10 to 20 degrees of frost on the docks, is almost enough to kill an old fellow like me.'

The *Durham* preparation in January 1876 reinforced that truth to the 64-year-old. 'So severe was the frost and so cold the north-east wind blowing through the shed that, before I had finished packing, many of the eggs were frozen and killed. The moss also kept freezing and had to be taken to the galley fire on board the *Durham* to be thawed before I could use it. Never did I feel the cold so much.'

Despite a quick 63-day voyage to Melbourne, two-thirds of the eggs had perished and in unseasonably warm March weather hatchings proved 'a complete failure', save for five at Wilson's Ercildoune estate. Wilson blamed the losses on eggs being shipped immediately after impregnation, suggesting 'our English friends' could learn something from the French 'wait for the eye' method, as older embryo could better stand ship movement and rough handling. Until this approach was adopted 'no future shipments of English salmon can be sent with any hope of results ... anything like satisfactory'.

Youl's *Norfolk* and *Lincolnshire* shipments had already shown eyed embryo was not essential, but Wilson said his knowledge came from those involved in the American salmon rush. At the 1876 US Centennial Exhibition in Washington, Baird boasted of 'a great triumph for our Californian salmon' in Australia and New Zealand when 'so many pounds sterling had been expended in trying to introduce the Scotch salmon ... all the experiments in Australia and those in New Zealand had failed.'

It annoyed Youl that the views and reputations of 'republican' Americans outweighed 'loyal' Britons. Francis Francis protested on Youl's behalf to *Field*. Tasmania might be held 'as no part of Australia', he said, but 'fortunately, the fame of Mr Youl's splendid exploit does not depend on American laudation or American suppression.' Baird responded that while he had not specifically referenced Tasmania 'I have kept myself familiar with the efforts of Messrs Youl, Buckland and others to introduce fish into the Australian colonies and am fully aware of the success in Tasmania.' Given that, *Field's* editor noted it was 'strange' he 'should have passed over without remark the greatest piscicultural achievement of the age'.

But achievement continued to be shadowed by the 'if' word. 'If migratory salmon bred, as alleged, in Tasmanian waters, why has no one ever caught or seen a fish over the debatable borderline of eight, 10 or 12 pounds?', asked Anglo-Australian journalist-angler, William Senior, in the *Gentleman's Magazine*. 'It is heresy to suggest this doubt to a Tasmanian, but to this moment though there is every presumption in favour of the existence of the noble *Salmo salar* in Tasmania, there has been no proof positive such as would be required to satisfy a jury of experts.'

By now, Senior said, from the first salmon hatched in May 1864, there should be salmon in the Derwent of 60, 70, or 80 pounds (27 to 36 kilograms). Frustrated Hobart fishermen petitioned to have the whole Derwent estuary opened to netting. Allport fiercely objected, saying the resulting 'annihilation' would destroy 15 years of effort. The colony would be 'justly held up to the contempt and ridicule of the whole civilised world, first spending about £15,000 over the experiment (£1 million today, A\$3.3 million), and then in defiance of warning, sacrificing the whole at the dictation of half a dozen fishermen and poachers.'

Sir Robert Officer, in a Christmas 1877 letter, told Youl of what he saw as 'the crowning event in the lengthened history of our undertaking'. Salmon were returning to make their nests in the Plenty, and he had seen a female weighing 20 pounds after it had laid its spawn, and a male weighing 18 pounds. 'With the salmon breeding in the Plenty ... abundance of ova and young fish may hereafter be obtained for stocking speedily all the rivers in Tasmania.' He hoped to soon send a fish 'of sufficient size and condition ... to yourself, to whom we are so much indebted'.

Allport also assured Youl 'you may depend on my sending the first fish weighing anywhere over 20 pounds that I can lay my hands on straight to you'. He was prepared to pay 'a very fancy price' to secure any large specimen which appeared, but he died suddenly in late 1878. Youl did not enjoy an *Argus* obituary declaring it 'was to his (Allport's) ardour and perseverance that the success of the acclimatisation of salmon in the waters of the Derwent was chiefly due', and *Field* stating no one had done more to confer pisciculture benefits on all the colonies.

Youl wondered might be said of him when he passed, more so when Samuel Wilson published *The Californian Salmon: With an account of its introduction into Victoria*, quickly followed by *Salmon at the Antipodes: Being an Account of the Successful Introduction of Salmon and Trout into Australian Waters*. Wilson gave credit to Youl for his packing breakthrough, but declared that if salmon proved successful it would be a just reward 'to the writer for all his exertions'.

In reviews, some newspapers headlined their reports 'How the Salmon got to Australia', while the *Literary Examiner* said 'if any man is entitled to write about the introduction of the *Salmonidae* into Australasia, it is Sir Sanuel Wilsom, for he was amongst the earliest, the most enthusiastic and self-sacrificing of those who labored to bring the king of fish into the rivers of New Zealand, Tasmania and Australia.' The *Saturday Review* lauded a 'patriotic enthusiast ... sparing neither money nor trouble ... in a series of adventures which involved him in considerable hardships.'

Wilson's book re-opened old sores. Henry Moscrop wrote to the *Morning Post* and *Land and Water*, rejecting Wilson's credit to Youl for the discovery of the mode of packing ova in boxes of moss and charcoal, then placed amid ice. He reiterated that, as he told the *Times* in July 1863, it was his suggestion of packing the boxes in ice, which proved successful on *Beautiful Star* and was further confirmed by experiments he instigated, leading to the *Norfolk* success. While Youl 'assumes the credit ... I claim now, as I ever have done, the credit of having originated the idea and to be the inventor of the process by which this grand result has been obtained.'

Moscrop also told the Tasmanian and Victorian governments he had sent a draft of his *Times* letter to Youl and William Ramsbottom, and both had signed responses approving it, Youl adding it presented 'all the facts of which I am cognisant of'. Youl had also acknowledged receipt of a similar letter to the *London Review*, and 'during all this correspondence Mr Youl never denied to me the credit of having made the discovery'. Some years later when Moscrop met Youl, 'to my very great astonishment he claimed the credit of being the first to suggest he idea of retarding the development of ova, and the mode of packing it ... (he) utterly denied my account from beginning to end.'

Youl saw the Moscrop correspondence in the *Morning Post* as 'a vile attack ... artful and ingenious' but again rather than directly respond he turned to Lachlan Mackinnon, the former *Argus* co-owner who had been part of the early Colonies Association salmon efforts. From his estate at Duisdale, Isle of Skye, Mackinnon wrote to newspapers saying he had no wish to detract from Wilson's efforts in Victoria but any impression that he had been first to introduce salmon was erroneous. His was a subsequent and 'totally separate enterprise' from Youl's success.

Youl thanked Mackinnon for 'your manly defence of a credit which is justly my due, and which so many persons have laboured to rob me'. Without papers and correspondence which Mackinnon had retained 'it would have been hard to prove myself a truthful person'.

Youl did prosecute his 'truth', in a letter to Wilson, a copy of which was provided to the *Argus*, and Mackinnon. He said Moscrop was 'the third person who has appealed to me, as being the accoucheur who brought this little bantling into the world, to get a certificate from me which each of the three claims as his being alone the father. Unfortunately for them, and to their extreme disappointment, I have not been able to comply with their request, or with that of some others who have put forth a similar claim; knowing very well who the real originator of the packing of the ova in boxes was'.

Youl said 'the real parent ... to whom I had always given the credit' was 'Girley' (sic, the Frenchman Zepherin Gerbe) who 'showed him how fish ova packed in wet moss in earthenware jars were sent long journeys'. Gerbe didn't think the method would succeed on a long journey, 'but it gave me the outlines which led me to my having made the little box, packing it in moss bought, selected and washed by myself, the ova packed with my own hands, and place in the icehouse on board the *Beautiful Star*.

'There is no doubt with every disinterested person who took an interest in the subject at the time ... that it was entirely owing to my former experiences in the shipping of ova and building an icehouse in the *S.Curling* that I was able to work out (the French) suggestions so as to obtain success'.

With so many testimonies of 'my success', including French awards and 'scientific press', Youl felt he should have ignored Moscrop's letter 'with silent contempt' but the man had 'a craze' that as he had received a £12,000 reward (£ 1.25 million today, A$2.6 million) Moscrop was making claims on Victoria and Tasmania for his share. 'Be on your guard,' he told Wilson.

Youl also maintained to Mackinnon that while the French showed him their use of moss in packing, 'I never should have (ultimately) succeeded but from the experience I had gained in building the icehouse and fitting out the *S.Curling*.' While the shipment failed because of Alexander Black's poor management, it 'had the effect of rousing the governments of Victoria and Tasmania that the *Salmonidae* might be introduced into their rivers'. And Wilson's success with Californian salmon 'would have been impossible but for my demonstrating how the thing could be done'.

'When I look back,' he said, the wonder to my mind is how I was enabled to persevere for such a long period of years, from the discouragement I received from so many who understood the subject, and the abuse heaped upon me by the press and people of Hobart Town.

'For years I had to grope my way in the dark. There was nothing to guide me. No one before had made the attempt to use ice for the conveying of ova. I built the first icehouse for this purpose.'

Failures, he said, gave him the experience which finally enabled success with *Norfolk*, 'which astonished all the eminent scientific pisciculturists whom I had consulted, and if I had given it up before I had succeeded no one else could have done the work without going through all I had done ... it was simply because I worked out all the details myself that I did succeed.'

As for ultimate salmon verification, the question remained unanswered, although Dr Francis Day was adamant the British Museum housed one specimen, sent by Royal Society Museum curator, Thomas Roblin, which was labelled, 'and correctly so ... an undoubted salmon smolt'.

And Gunther had also somewhat softened his position. In a chapter on acclimatised species in *An Introduction to the Study of Fishes*, he referenced 'the most successful attempt of recent years is the acclimatisation of trout and sea trout, and probably also of the salmon, in Tasmania and New Zealand'. He fully accepted trout and sea trout had become acclimatised, but whether salmon would be 'equally and permanently successful remains to be seen'. Gunther still maintained true salmon was a Northern Hemisphere fish, but said characters and habits of species could be affected by transfer to distant parts of the globe, and 'numerous cross breeds have been introduced into and reared in Tasmania ... (interfering) with ... pure breeds.' Identification within 'the labyrinth of confusing variations' was of 'considerable difficulty' with scope for 'great diversity of opinion'.

While salmon continued to bewitch, time marched on. Amid 1880 Christmas festivities, Youl's wife Eliza passed suddenly at 63, a victim of heart disease. So too Frank Buckland, of lung inflammation, but not before he used the final months of his life to stake his claim to fame. Finishing *Natural History of British Fishes* two days before he died, Buckland laid claim to the trout abundance in Australia and New Zealand: 'I may fairly say that these colonies owe the existence and almost abundance of trout at the Antipodes to myself.' But he lamented, extravagantly, that while trout was 'almost as important to the colonies as the introduction of sheep', his and Francis Francis's ova contribution had not received public acknowledgement.

Worried 'it will be forgotten', he asked a friend, prominent Melbourne surgeon Dr Robert Stirling, to raise the matter in Australia. The Victorian Government was more focused on a 'dead or alive' pursuit of the famous Ned Kelly gang, but the acting Chief Secretary responded: 'I think the time has arrived when it should be placed on official record that the honour of initiating such acclimatisation in Australia is certainly due in the largest degree to yourself and Mr Francis Francis who jointly presented to the Tasmanian government the boxes of trout ova in 1864 for the purposes of introducing that fish into Tasmanian waters.' With Victorian rivers now well stocked, the colony was indebted to him for 'demonstrating the practicability of conveying the fish indigenous to one hemisphere to another'. (This despite, more than a decade earlier, the Acclimatisation Society of Victoria awarding a medal to Youl for his introduction of salmon and trout ova, stating the 'deep obligation' of all the colonies for his 'persevering, enlightened and patriotic efforts').

In his final words, Buckland described trout in the Antipodes as 'the greatest feat of pisciculture of modern times', but recognised 'the energy and perseverance of my friend Mr Youl'. As for salmon, Buckland had previously said if salmon was eventually acclimatised the Antipodean colonies 'will hereafter look upon Mr Youl as one of the great benefactors of his age ... (but) whether it will be possible to establish salmon at the Antipodes as yet is an open question.'

Understanding that success always had many fathers, and anxious about his legacy and reputation, Youl looked forward to his account being told by Arthur Nichols, a fellow of the Royal Geographic Society, to whom he had provided his entire collection of newspaper and journal clippings, papers and memoranda, maintained with the help of his daughters. His anticipation about Nichols' forthcoming book grew with Nichol's account in *Chambers's Journal*, in which he said: 'Here was a problem apparently beyond the power of human skills to solve ... yet within 12 years from the date of the first attempt salmon were seen to be seen swimming in a Tasmanian river!'

The 'brilliant success' meant salmon, 'salmon trout', and brown trout ova had been widely distributed and 'cannot fail to establish themselves in the course of time', Nichols wrote.

But no man can count on time, nor salmon.

24

ANOTHER SNAG

Tired of the endless 'where's the salmon?' question, Youl became more convinced local fishermen were culpable.

He and Morton Allport had long suspected that fishermen had 'all along endeavoured to destroy evidence' because salmon protection regulations prevented them from all-year netting of the river for native fish, or because they believed the salmon reward would inevitably increase. According to Arthur Nichols, 'a little judicious bribery revealed an organised system of murder of all strangers found in the river.'

Youl wrote to the *Times* about receiving numerous accounts of Derwent fishermen netting 'large numbers' of salmon grilse weighing from a pound to three and a half pounds and selling them for 6 pennies a pound. He held 'great fears' that the destruction of so many young fish evidently returning from the sea to spawn threatened 'the complete success of the experiment after all the expense and care ... in successfully placing the fish in the Derwent'.

Field shared the concern: 'We have no doubt in our own minds that the salmon has been satisfactorily acclimatised in the Derwent. Whether it will stand the operations carried on by the net fishers at the mouth is another question.' Shoals of migrating smolt were easily swept up and two or three years of such work by 'wretched poaching idiots ... (who) would render useless all the trouble and expense gone to'.

Youl was also anxious for the findings of a Royal Commission into all the colony's fisheries. It was accepted that certain types of *Salmonidae* had clearly been acclimatised, but the commission probed 'what types?', or 'what type of types?'

There was no doubt about trout. Rising ichthyologist Robert Johnston, who migrated to Tasmania from Scotland, said the brown trout success was now 'famous'. 'It is no small credit to Tasmania that she is the first colony ... which has succeeded in the remarkable achievement of stocking her waters with European fishes.'

But evidence of salmon success was less clear. Smolts 'dressed in their bright migratory scales' had for years come down a water-race at the Ponds 'in millions' to enter the Plenty, but 20 years on all that was known, he said, was that the island had 'fine' non-migratory brown trout and 'splendid' sea-going migratory salmonids. Johnston first thought the waters could contain up to seven varieties: 1. Brown trout; 2. Sea trout; 3. Salmon trout; 4. Atlantic salmon; 5. all of these in variable numbers; 6. a hybrid taking in some characteristics of the four; or 7. one of the four modified in a new environment.

Numerous fish did not fall within exact classifications of any *Salmonidae*, yet possessed distinguishing features of more than one of the salmon, sea trout and trout. It was 'satisfactory' that some specimens sent to England had been confirmed as having 'all the characteristics of true salmon',, and so the successful acclimatisation might still be 'a matter of time'. But 'we might, however, have expected that ere this a large specimen of *salar* would have been captured.'

The Commission and Johnston lauded acclimatisation pioneers, especially Youl, for 'enlightened and far-seeing views', and with Scotland and Ireland fisheries now respectively worth £750,000 (£76 million today, A$159 million) and £500,000 (£51 million today, A$107 million) annually, 'it is to be regretted that their praiseworthy efforts have not been more adequately followed'. Youl would have enjoyed the sentiment but chaffed at the report stating that it was to the late Allport, whose son Curzon was one of the Royal Commissioners, that 'the success of the (*Salmonidae*) experiment in Tasmania is largely due'.

In an accompanying report, *General and Critical Observations on the Fishes of Tasmania*, Robert Johnston pointed out that experts like Dr Gunther in England and Professor Frederick McCoy in Melbourne had the disadvantage of trying to determine 'salmon' from single specimens, and sometimes not even a full specimen. In 1877 McCoy, the Irish-born natural science professor and Melbourne Museum director, judged two fish heads sent by the mayor of Hobart to be 'true salmon', then a few months later judged an entire fish at the Victoria Club a sea trout, and then another fish head to have the characteristics distinguishing salmon from the sea trout and other species.

'I am not surprised,' Johnston said, 'that ... at different times they have pronounced certain individuals' to be salmon, sea trout, or a hybrid. Odd specimens cannot determine the curve of variability, nor can they determine whether the ... fish, so differently named, were not after all the progeny of the same parents.'

While some still saw salmon as 'fool's gold', the Royal Commission saw enough encouragement from the success of trout and 'salmon trout', and some 'true salmon' affirmations by leading scientists, to call for annual shipments 'until the introduction of this species is beyond the region of doubt'. The idea of fresh shipments, perhaps for six years, to ensure sufficient numbers to 'silence all the doubters', gained momentum. Having come so far for so long it seemed the 'greatest effort' was justified. And Tasmanians did not want to be defeated by America's Fish Commission gifting salmon ova to New Zealand, Victoria or the Zoological Society of Australia for attempts in New South Wales.

To assess the cost and viability, the Salmon Commission turned to one its own Irish-born Dr James Agnew, in London on a 15-month sojourn. The Commission was confident the government would fund another shipment as shipment of ova 'under the influence of cold' had been established, steamers with new refrigeration compartments were carrying Australian and New Zealand beef and lamb in as little as 30 days. Success was now 'almost sure ... without the great trouble, expense and delay' of previous arrangements.

Agnew first called on Youl, rating him the only individual in London who would devote the necessary personal and intelligent attention to detail. They met several times at the Royal Colonial Institute and over dinner at Youl's Waratah House. Youl was pleased Tasmania had decided to end a 15-year hiatus, during which time he had transported more than 500,000 ova in a bid to meet New Zealand's appetite. Despite feeling his age, Tasmanian salmon remained unfinished business, and he agreed to 'give the experiment all the aid in my power'.

Ireland's fishery inspector Thomas Brady, seen by Tasmanians as the man with the greatest knowledge of Ireland's salmon and its supplies to English markets, offered to freely provide salmon ova from Irish rivers for three years. It led the Tasmanian Parliament to provide another £500 to finally 'set at rest' the interminable controversy.

Agnew was given 'uncontrolled direction in the matter'. He tried to persuade Youl about using a fast, refrigerated steamer to Melbourne, and on-shipment to Hobart. But he wasn't the first to find Youl not easily persuaded. Refrigeration was new and Youl remained 'strong' in concerns about steamer vibration. 'I doubt if he will alter his opinion,' Agnew reported. 'He was pretty positive on the point.'

Agnew then suggested three trial methods. One batch of moss-iceboxes to be packed in blankets and enveloped in a thick covering of dry straw and planted deep in non-conducting snow produced by a steamer's refrigeration. A second batch to be planted amid two-foot-thick blocks of ice just before the refrigeration compartment reached a freezing point. And a third trial of boxes of ova covered with straw, or any better non-conducting material, and perhaps canvas, left exposed in the chamber.

Youl insisted on discussing the ideas with Professor Thomas Huxley, the leading scientific voice at the time, who had been drawn to the salmon quest. Huxley presciently envisaged that steam and refrigeration

MR. T. F. BRADY, INSPECTOR OF IRISH FISHERIES.

Thomas Brady, Irish fishery commissioner.

developments would one day 'make it possible for us to draw upon the whole world for our supplies of fresh fish'. He could see his son or grandson 'when he goes to buy fish ... offered his choice between a fresh salmon from Ontario and another from Tasmania.'

But, according to Youl, the scientist 'entirely concurs in my opinion' that the existing refrigerators in the Orient steamers 'would not be likely to convey the ova live to Australia'. However, he was investigating the viability of using an outer chamber in which the temperature was perhaps 44°F (6°C).

At the 1881 annual Christmas dinner of 'Tasmanians and friends' at the newly opened Anderton's Hotel in Fleet Street, organiser Seymour Bennett toasted Youl as one who had 'achieved immortality', then roasted *Field* and others for suggesting he had not been adequately recognised by Tasmania. Youl was a 'household name' in Tasmania, Bennett declared, 'and it is almost impossible for his name ever to be forgotten there, and I think that must be one of the greatest rewards that a man can earn, to have his name remembered with gratitude by a people ... I am sure he looks for no other recognition than that. That his efforts have been recognised officially we all know and see and are proud of.'

After loud cheering, Youl was too polite to air any disgruntlement about the amount of recognition or being rewarded with something more tangible, perhaps the 5000 acres (2000 hectares) he had mentioned when his quest began. He merely said, 'I endeavoured for many years to do the best I could for that little colony.'

But with no suitable ship sailing direct to Hobart, and the spawning season underway, another mission was postponed. Brady lamented the 'mortification ... a whole year must now be lost.' Agnew, needing to return to Tasmania, persuaded the government to depute his authority to a committee of management: Youl and Brady in charge of the ova, and expenditure in the hands of bank director-merchant Richard Philpott, already representing the colony at the International Fisheries Exhibition.

The delay was regretted, but Youl, now 70, was favoured on other fronts. In 1882, less than two years after Eliza's death, he wed German-born Charlotte Robinson, a 48-year-old widow who lived nearby in Clapham Park. They married at Her Britannic Majesty's Legation in Brussels, Belgium, where Charlotte's family lived, before returning to Waratah House to live comfortably with two maids and a domestic cook.

And Arthur Nichols' book, *The Acclimatisation of the Salmonidae at the Antipodes, its History and Results* was finally released. Nichols admitted he had initially been sceptical of Youl's quest when he heard how Robert Ramsbottom once dropped a soda water bottle of ova and none survived. 'The writer was prepared to believe that the task of conveying them to Australia was hopeless.' But since then, Youl's 'persevering, enlightened and patriotic efforts (had led to) ... a singular achievement in pisciculture'.

'He who succeeded in making two blades of grass grow where but one grew before has been canonised as the greatest benefactor of mankind,' Nichols said. 'But surely he who achieves the more difficult task of transplanting an animal from one hemisphere to another and peopling a barren river with a noble species of fish, should not pass unnoticed.'

But, Nichols wrote, when Youl ultimately triumphed he had to fight off others claiming to be 'the originators' of the transport plan. Or it had been claimed for them by others, giving some an 'undeserved honour'.

Nichols wrote that Buckland had repeatedly claimed success 'and he has not been careful to disavow the honour at all times and in all places'. While he had done much good service in fish culture, any impression 'he was the author of the acclimatisation of salmon at the Antipodes must be dispersed'.

Another target was Dr John Gray, keeper of zoology at the British Museum and, like close associate Gunther, one 'very incredulous' of the salmon quest. Gray had long felt Youl poo-poohed his mid-1850s suggestion to look at how trout 'packed in ice' had been sent to Dr John Davy for his Charles Darwin experiments. Gray revisited it again five years later in the *Athenaeum*, saying 'the Australians ... laughed at me and attempted to ridicule the idea'. He declared 'all that Australia has done in this business ... has been an utter failure and very expensive'.

To kill off Gray's campaign, Youl turned to Henry Watts, a former *Argus* editor back in London writing editorials for the *Standard*, and later contributing to *Encyclopedia Britannica*. After Youl helped 'refresh my memory', Watts said he was able to 'affirm most positively' that Davy had nothing to do with the plan adopted by Youl, who had 'never met him in his life' and derived no hint from him or any other scientific man in England.

Whatever the precise 'share of the discovery' of others, Watts said, 'the credit of having discovered a mode of packing and transporting the ova to Australia ... belongs chiefly to Mr James Youl ... the success which was ultimately achieved overcoming opposition by all the savants and experts – in conception and execution entirely the work of Mr Youl.'

It was unfair, he said, for an unscientific man like Youl to be denied 'the sole honour of having introduced salmon into Australia' when men of science and experts 'gave us no help whatever in the undertaking ... did not little to thwart it by their sneers and objections ... sinister prophecies.'

When Nichols' book ignored him, Henry Moscrop revisited his claim to be the inspiration of the *Beautiful Star* box trial and subsequent experimental proof in the Wenham Lake ice facility. He expressed his 'astonishment' at the extent of Youl claiming the credit, telling the *Morning Post* and *Land and Water* that without his suggestion and discovery,

Youl would probably have continued to waste money pursuing 'his own impracticable notions'. Moscrop's claim that the experiments were 'at my suggestion and my expense' was supported by Youl's assistant, Thomas Johnson, who told the British Association of experiments which 'tested the value of a suggestion by Mr Moscrop (of) burying the ova in ice'.

Notwithstanding the contested versions, *Field* described Nichols' book on Youl's quest as a relief. Amid 'worldwide interest' it had long insisted 'that Mr A, B, C or D had nothing whatever to do with the transport of ova of *Salmonidae* to Australia'. Injudicious friends posing as public benefactors were 'simply depriving those who really did carry out the work at such personal labour and trouble of the just credit.' Now, 'fictitious' claims would no longer be countenanced.

'Fictitious' claims were perhaps behind him, but Youl had to await the next spawning season for any advance on what some still saw as a salmon fiction. 'What became of them?' the *Tasmanian News* asked. 'Heaven only knows.' And Youl lost his key ally, Robert Ramsbottom: on an 1884 Christmas visit to his son Robert Jr in Manchester, 'the eminent pisciculturist of Clitheroe' succumbed to bronchitis or lung disease. Perhaps, like his late son William, another victim of too many freezing days and nights in pursuit of salmon.

It was sad news on his birthday, but Youl perhaps thought another mission might suitably honour his old friend. Then even greater motivation came in a letter from his son Charles at Symmons Plains. It contained a salmon bombshell: one of Britain's most respected fishing men, who had become the Tasmanian Government's first fishery inspector, had declared a 'mistake' had been made.

The original 'salmon' eggs, from the *Norfolk* expedition, were not salmon.

25

FICTION & FRICTION

'Impossible!' Youl fumed. 'I employed the best and most trustworthy men in all England to collect the ova for me, packed all but a few boxes myself, and am certain they were true salmon eggs.' He and other men of repute had also retained and hatched hundreds of eggs from the same consignments, 'and all agree they were *Salmo salar* ... there was no mistake.'

Ironically, it was a desire to avoid mistakes which led to Tasmania appointing its first fishery inspector. The Royal Commission into the island's fisheries wanted to avoid a repeat of the ignorance, greed, and poor management which had decimated the island's former whale and oyster industries. It recommended more professional and unified management of salmon, other fish and seafood, including oysters, under the control of a central board.

Salmon Commissioners pressed to remain the czars of *Salmonidae*. Since William Ramsbottom's death 15 years before, they had pressed

William Saville-Kent, Tasmania's first superintendent of fisheries, 1997. Photogravure by Waterlow and Sons, from negative by Maull and Fox, published in *The Naturalist* in Australia by W Saville-Kent, Chapman and Hall, London.

for an educated man to supervise the Salmon Ponds, exercise 'general control' over fisheries, and assist private fish and oyster farming. And everything to fall within the remit of a reconstituted Commission.

Without fully resolving the control issue, the government agreed the much-admired and newly knighted Irishman Thomas Brady be asked to help find a suitable 'superintendent and inspector of fisheries'. The terms were a salary of £350 (£37,000 today, A$77,000) a year for three years, and a first-class passage to Tasmania, although 'not for the wife or family should he be a married man'.

Brady wasn't sure he could find a man who understood both pisciculture and ostreiculture, but he would try 'and should I succeed will make him an offer', with government agents in London authorised to endorse 'anything I may do'.

Brady turned to Professor Thomas Huxley, Her Majesty's new inspector of fisheries following the death of Frank Buckland, who proposed 38-year-old naturalist-author William Saville-Kent, formerly

William Kent, who he had mentored. He had impressed Huxley with his labelling of Buckland's fish collection when it was bequeathed to the British Museum, and preparing a handbook for the Great International Exhibition in London in 1883, which helped establish the global nature of fisheries. (Youl entered his own exhibit: a copy of Arthur Nichols' book detailing his 'triumph' and a 'facsimile of boxes and mode of packing salmon ova sent to the Antipodes'.)

Saville-Kent was attracted by the prospect of creating his own research base in the Southern Hemisphere. It also gave him a chance to get away from his past. As teenagers, Kent and his sister were questioned about the brutal 1860 murder of their three-year-old half-brother Saville. She was convicted and jailed, and he took up the hyphenated surname of Saville-Kent.

Brady told the commissioners he had found their man: a 'distinguished scientist, very highly recommended by one of the first scientific men in England, Professor Huxley'. He would sail in about two months.

Youl hoped anyone chosen by Brady, known in Ireland as 'the fisherman's friend', would deliver the long-awaited verification of true salmon. The pair were preparing for a fresh ova batch from Ireland, but Brady's workload forced him to rely on his son Herbert, on leave from a diplomatic career in China, to collect 120,000 ova from the Erne, Roe and Blackwater rivers. In sealed bottles of ova placed inside boxes in a wooden crate suspended by Indian rubber rings to keep movement to a minimum, the ova were trained to Dublin's Amiens Street station and shipped to the Wenham Lake ice stores in London.

Youl spent more freezing days packing ova, this time on a new three-masted steamer, *Abington*, sailing direct to Hobart. It was a struggle. 'They would adhere together as if glued ... several persons say it is owing to the eggs being pressed out of the female fish before they are ripe.' He 'took all the pains' he could but after a protracted seven-week voyage, with a drainage defect in the icehouse causing ova boxes to float and be knocked about in rolling seas, saturated and rotting moss and dead ova became a putrid pulp. When only 1825 fry were hatched, Brady promised to personally oversee the next collection and delivery and send two or three hundred thousand eggs.

The Salmon Commission was pleased to receive from Brady and Youl two preserved adult females and a male to help resolve public disputes about species identity, but it faced a bigger salmon war. Shortly after Saville-Kent's arrival in July 1884, parliament passed an Act legitimising him as 'superintendent and inspector of fisheries', under the Chief Secretary, to act as scientific adviser in 'all matters relating to the regulation and development of fisheries'.

The Salmon Commission protested about the 'danger of misunderstandings and want of harmony'. This quickly became evident. Just two months after arriving Saville-Kent reported to the Chief Secretary his shocking declaration: There were 'no true salmon' established. 'No fish that I have so far seen possesses the full characteristics (of salmon)' he said, and after 'careful consideration' of all the evidence, he could come to no other conclusion other than 'by some accident the eggs of the salmon trout, instead of those of the salmon, were in the first place imported from England'.

Saville-Kent wrote to *Nature* journal that many of the fine *Salmonidae* in the Derwent were a variety of brown trout, grown to a large size due the 'superabundant' food supply. He believed the 'rapidly accumulating' evidence was that the common brown trout and migratory 'salmon trout' were extreme racial variations of the same species, living in either fresh, brackish or salt water, and that the two could freely interbreed and produce fertile progeny.

After 20 years, he said, it was 'scarcely possible' that only young salmon should have fallen a prey to other predators, but by now there ought to be evidence of salmon weighing 40 or even 60 pounds (18 to 27 kilograms). Nevertheless, the island had a 'world-famous trout hatchery' on the Plenty River, and he remained optimistic about the salmon prospects but urged the commission to ensure it had 'well authenticated' salmon to turn into rivers devoid of any trout after it 'most undesirably' released shoals of brown trout 'into every accessible stream in the colony without thought for the future'.

Saville-Kent also wanted a 20-year ban on Derwent netting be lifted for three months because, he argued, there were no salmon in the estuary to protect and the 'pseudo-protection of the rivers ... by an association having no responsibility and legal status (is) a mere farce'. And he wanted to utilise part of the Salmon Ponds for breeding freshwater grayling or herring to overcome decimated stocks in the Derwent.

The Commissioners saw his comments as 'calculated to create an antagonistic feeling'. A *Mercury* contributor went further and wondered 'they did not have him hung at once'. And there were mutterings about reputations being threatened by an outsider who had once been questioned about a child's murder.

Saville-Kent complained the commissioners took every opportunity 'behind my back' to thwart his work and damage his reputation. He was told the majority had decided there would be no concessions. 'Either I or they are to be smashed.'

In London, Dr Francis Day consoled Youl by saying that while Saville-Kent may not have seen a true salmon that 'does not absolutely prove their absence'.

Youl and Brady pressed on, trying a small experiment on *Tainui*, sailing to Tasmania. Thomas Johnson built a small apparatus containing six perforated trays for ova and moss, interspersed with trays of ice water from a refrigerator. A Brady friend on board was entrusted with regular replenishment of the trays. But after the ova arrived in March 1885, only 50 fry were produced. It was unclear whether the issue was unsuccessful fertilisation or difficulties maintaining the temperature, but two months later the Canterbury Acclimatisation Society in New Zealand reported much greater success with a similar trial with 200,000 ova on a new refrigeration steamer, the 4474-ton (4058-tonne) *Kaikoura*.

Under pressure to deliver the goods, Brady secured 150,000 ova for Youl from the Erne and Blackwater rivers, travelling night and day on four trips in 'a whole first-class carriage filled with my cases, all swung in such a way there could be no addling or concussion.'

In another chapter of what he called a 'labour of love', Youl packed 101 boxes, including a batch of 10,000 more developed 'eyed' ova, as the French had long advocated, on a new steamer, *Yeoman*. With his fastest ship, the largest number of eggs, and a refrigerated room keeping

Steamship *Kaikoura*. Courtesy NZ Ship and Marine Society.

the temperature just two degrees above freezing, Youl anticipated his biggest success.

After a 66-day voyage the 150,000 ova landed in May 1885 led to 36,000 fry, much more than *Norfolk* and *Lincolnshire* outcomes. While the Commissioners and Youl still favoured the icehouse approach, as it avoided the risk of refrigeration breakdown and the cost of an attendant, Dr James Agnew, now back in Tasmania and about to become premier for a brief period, was taken with the *Yeoman* and *Kaikoura* successes. He offered to personally meet the £500 (£55,000 today, A$114,000) cost of a *Kaikoura* shipment of 'eyed' ova only, as it was seen to be more successful, and for Brady to deliver the ova to Youl to pack and then accompany it to Tasmania. The plan was backed by Governor Sir Robert Hamilton, who had befriended Brady as under-secretary in Ireland before his pro-home

Barque *Yeoman* carried 150,000 eggs for Youl. Courtesy Library of South Australia.

rule sentiment led to an 'Irish rise' or 'promoval' – the removal of one who couldn't be kicked downstairs – to faraway Tasmania.

Meanwhile, to Youl's grief, Saville-Kent's continued 'no salmon' declarations caused *Field* to reverse its supportive position and declare 'we must accept his testimony as a serious denial of all previous statements as to salmon caught in the river Derwent'. Youl hoped the impasse would be resolved when Brady delivered his judgment, along with 400,000 ova on *Kaikoura*, almost the total of six previous voyages. The *Kaikoura* voyage was groundbreaking for another reason: it brought out 'the English footballers' on the first major tour by a European rugby union team.

But salmon remained the main game. The press welcomed Brady as the 'grand old man' of salmon and fisheries, Hamilton accommodated him at Government House, and crowds cheered as he sailed up the Derwent to the Salmon Ponds. The *Tasmanian News* declared 'The Salmon Problem Solved', as Brady would ensure a golden harvest was not to be ruined by 'wretched inefficiency and sycophantic patronage', and the public looked forward to him showering 'sparks from the anvil of his brain'.

Brady's sparks immediately showered at the Plenty Ponds, where he declared he now considered 'no difficulty in carrying live fish through the tropics'. He had experimented on *Kaikoura* with seven one-year-old fry in glass water jars, which he continually refreshed and fed with 'worms' made of flour and water. He had kept them alive until they were six degrees south of the equator. Under certain conditions, he said, it was 'quite possible to ship fish alive in London and land them in that condition in Australia'.

A few days later he created a bigger sensation at a Royal Society event, keenly hosted by Governor Hamilton, who considered he had recently caught the truest and largest salmon in the Huon River south of Hobart. He sent a photograph of the 'grand fish', a 29 pounder (13 kilograms), to *Field* for its judgment. William Senior, an Anglo-Australian journalist and angler writing as 'Red Spinner', saw enough to have doubts. It had been 'vaguely described' as 'a splendid specimen of *Salmonidae*' and lacked examination by Saville-Kent. 'That it was a salmon I do not believe ... that it was a silvery fish that could readily pass for *Salmo salar* I understand.'

Saville-Kent had looked at the governor's catch but felt himself unable 'conscientiously to pronounce the verdict that was expected', while Robert Johnston thought it corresponded more to 'salmon trout' than salmon. The island waited anxiously for Brady's judgment.

It was 'a true salmon', he declared. 'No practical man who would see the fish would ever think of calling it anything but a salmon. No scientist would consider or talk of the fish we have in Tasmania in any other way but as a salmon.'

The pronouncement was received 'almost with open-mouthed astonishment', after Tasmanians had become so accustomed, the *Mercury* said, to scientists like Saville-Kent and others declaring that captured *Salmonidae* 'fail to fulfil this or that test supposed to mark the true *Salmo salar*'.

Brady was not scientifically trained, but well experienced and firm. Or firm-ish: 'Whether it be the true *Salmo salar* or not it is, at any rate a fish which would be considered and treated as a salmon in salmon countries ... sold and purchased as such. If the colonists of Tasmania seek for more than Ireland, which now exports salmon to the amount of over £600,000-worth (£65 million today, A$138 million) annually, he could not help saying that ... they are hard to please and ought to go without them.'

Even 'the best scientists in England are at sixes and sevens on the subject', he admitted. This was evident a few years before, when he received three fish sent by the Salmon Commission. He thought one a true salmon, one not, one doubtful. A prominent Dublin ichthyologist concurred, before asking where the fish had come from. Told Tasmania, he then told Brady none were salmon, before conceding 'you may take it to the six best scientists in England and you will get six different opinions.'

Brady's opinion that the Governor's Huon catch was true salmon was accepted, notwithstanding that six years earlier the Royal Commission affirmed that salmon was released only in the Derwent, not the Huon, which had received 'salmon trout'.

Premier Phillip Fysh quickly sent a dispatch to Youl, now the colony's acting Agent General. 'You will be gratified to learn that the large fish caught by his Excellency the Governor at the Huon has been pronounced by Sir Thomas Brady to be a true salmon.' Youl immediately wrote to the *Times*, saying 'it shows that the *Salmo salar* has been successfully introduced'.

Field did not question Brady's pronouncement, but cautioned it was based on his experience as 'a practical man'. Numerous Tasmanian specimens 'no doubt look like salmon; all eat as salmon' but that did not make them indisputable salmon.

Half a century on from salmon identity wars in Scotland, Tasmania had a fish that looked like salmon and tasted like salmon, one that Brady, 'who probably knows more practically about the salmon than any living man', said was caught, sold and eaten in Ireland and England as salmon.

The fact of fish sold 'as salmon' in the English market was enough to please most locals, and it was seemingly justified under the English Salmon Fisheries Act, 1861, which said all migratory *Salmonidae*, including 'sea trout and brown trout', were 'salmon'. But Saville-Kent, struggling in the salmon and political currents, said if sportsmen visited Tasmania with the express wish for salmon fishing as they understood it in Britain, 'they are undoubtedly doomed to grievous disappointment'.

He achieved much in his two years as Australia's first professional fishery scientist, but Saville-Kent's stance on the salmon quest upset those with more friends. Asked by the government for his advice about management of salmon and all fish, Brady negatively critiqued Saville-Kent, but noted others also saw the Salmon Commissioners as 'amateurs'.

The Fysh Government resolved that a new 23-member Fisheries Board would be responsible for 'the general superintendence, management, and protection of the Fisheries in the Colony', embracing shellfish, oysters, and all fish, including salmon. Saville-Kent, visiting Melbourne, received a telegram advising his services would not be required beyond year's end. He was 'sacrificed to the petty jealousy of a clique', according to Launceston's *Examiner*. The Salmon Commissioners formally resigned after 26 years, but the new board included Commissioner Matthew Seal as chairman, retained Philip Seager as secretary, and others received lifetime appointments.

Saville-Kent felt too many who 'not having the courage of their convictions, have acquiesced and encouraged Tasmanians' in the belief that their superb trout were no less than 'lordly salmon'. Nevertheless, he generously praised the commissioners' 'indefatigable perseverance' and hoped Queen Victoria's golden jubilee might see the christening of a 'jubilee salmon'. If human skill could not overcome some 'inflexible law of nature' with salmon, he said, the colony still had varieties of the allied and more plastic forms of sea trout and brown trout, 'if not an equal ... a very substantial benefit'.

His final salmon take was that Tasmania's sea waters were perhaps 10°F degrees warmer than British seas and corresponded more with the Mediterranean. The millions of sensitive salmon fry and smolts entering the water would have 'wandered away, presumably in the direct of the Antarctic Ocean in search of cold water'.

While Youl wondered if he was fated to never have an unambiguous conclusion, it didn't fuss the *Sydney Morning Herald*, which asked: 'What more do we want? A rose by any other name?' Especially when 'experts' in London were not just 'at sixes and sevens' but had seemingly given up. In his 'labyrinth' of possibilities, Dr Gunther now said it was 'an impossibility' to identify artificially reared fish or their offspring, so would no longer attempt to do so without knowing their origin. Occasional fish from unfamiliar waters were even more problematic. Dr Day said many accepted theories about identity had 'failed in the proof', with former classifications abandoned and new 'keys' shown to be inconclusive. Professor Huxley admitted 'our knowledge is still far less exact than it ought to be', and salmon catch 'registrar' Henry Ffennell said 'it cannot be denied that we are profoundly ignorant on many points'.

The *Herald* said Tasmania's specimens did not align perfectly with any Northern Hemisphere salmonids but were 'so like the real thing that it is hard for the best-trained experts to distinguish them' that Tasmanians 'have no reason to complain'.

But Tasmanians had been pursuing true salmon for nearly half a century. Originally a 'mad' quest to many, it was now just maddening that the result was known by any name other than salmon. Fish identity was so divisive, one wrote, 'in some houses I would as soon admit an intimacy with the sea serpent as that I believed we have real salmon, and in others would not deem it safe to express myself contrariwise. There is a dread insecurity hovering over this question ... in uncertain company'.

Two decades after the first salmon fry entered the Plenty and Derwent, no good-sized fish completely aligned with Northern Hemisphere classifications. Salmonids commonly caught in the Derwent came close to the smolt and grilse form, but with variations often corresponding more closely with varieties of sea trout.

To fishermen, fishmongers and the public, a fish that looked like salmon and tasted like salmon was seen through salmon-coloured glasses. As Johnston saw it, 'in nine times out of 10 the silvery form, the colour of the flesh and the size alone determine their opinion, and all such forms are pronounced and sold as *Salmo salar*.'

That was enough for many, but not Johnston, who also felt waiting for London's anointment of 'true salmon' was a waste of time. The odd specimen sent back had only added confusion, with different specimens doubtfully pronounced as salmon, trout or sea trout, or some hybrid, without any detailed reasons for such conflicting conclusions.

'Authoritative opinions of this kind are worse than useless as we do not know the points of evidence upon which the separate opinions were based.' European experts had no knowledge of local conditions, so their opinions were 'purely rough guesses' without local study. Odd specimens could not allow them to determine the curve of variability, nor whether fishes differently named were not the progeny of the same parents. Variability led Gunther to assign *Salmo* names to specimens with key salmon and sea trout characteristics but also distinctive localised features, leading Dr Francis Day to protest that the profusion of salmon and trout names simply added confusion, when there were, in his view, merely many variations of the same fish. Dr George Boulenger, a leading zoologist who worked with Gunther at the British Museum, said it would be much better to 'describe the local forms as varieties instead of multiplying the species.'

Johnston, as much as Youl, wanted answers. If the Derwent specimens most closely corresponded with sea trout, when the only known shipment of such ova on the *Lincolnshire* in 1866 led to only 496 liberated hatchlings, could one really assume all the varieties common in local waters were descendants of those 496? And if so, what became of 37,824 'declared' salmon released over the years?

If the salmon were truly 'lost', Johnston saw the most likely explanatory theories as hybridism, extinction or exodus.

The hybrid theory was predicated on Youl's ova not being true *Salmo salar*. As Saville-Kent asserted, by some mistake the ova were from hybrid parents or a cross-fertilisation of *salar* and *trutta*. It was accepted that various species of *Salmonidae* interbred and perpetuated overlapping varieties, but Johnston felt Gunther's apparent conviction that only hybrid forms had ever been introduced was 'without justification'. With hundreds of thousands of eggs obtained for Youl by experienced men from distant salmon rivers in England, Wales, Ireland and Scotland, it was possible 'one or two mistakes' might have been made. But it was 'preposterous' to suggest so many experienced men had made the same mistake each time in different districts. It was 'untenable' to suggest this of men like Robert Ramsbottom, Thomas Brady, Frank Buckland and Francis Francis. Be it a fantastical conspiracy or coincidence, Johnston dismissed the theory.

The extinction theory, more commonly held, was based on the 'fatal shore' thinking that warmer water, different food, and Antipodean enemies were too adverse. Saville-Kent felt warner water was the most likely explanation, and Johnston initially thought it conceivable, but he had seen that even in warm shallow pond waters it had been proven *Salmo salar* could survive for many years and even breed. And Tasmania's food was rich and varied. As to rapacious foes, the number of migratory salmonids in the Derwent suggested little fear of the European type of *salar* having less chance of escape. With 97% designated salmon hatchlings against only two per cent sea trout and less than one per cent river trout, the numbers just did not add up. Johnston concluded the theory 'unsatisfactory if not untenable'.

The exodus theory, the longest held, was that the warm southern latitude water had not killed the salmon, but driven them towards colder waters. Saville-Kent suggested any liberated salmon may have migrated 'towards the south coast of New Zealand, or to Patagonia, or to the Antarctic icefields'. Or following some cold ocean current they might reappear on the shores of Japan or North-East Asia. But Johnston considered it inconceivable that some hereditary instinct might lead a Tasmanian salmonid northward, as trying to pierce equatorial latitudes would be like 'jumping out of the frying pan into the fire'. And if they did attempt 'this strange freak' of ancestral instinct they would somehow seek the Atlantic toward Britain, 'not in the opposite direction of Japan'. And there was no evidence that the temperature of open waters and rivers varied materially to southern portions of Ireland and England, and salmon hatched in colder New Zealand had also been 'lost'.

Johnston was now thinking less 'where's the salmon?' and more 'what's the salmon?'

He noted that Gunther and other scientists in Britain admitted the different climate, water and food in the Southern Hemisphere could have unknown impacts on various characteristics of transplanted fish. They also admitted imperfect or inconsistent characteristics in some local identifications. And equally there was no proof characteristics were not obliterated or modified in a new environment. As the chief inspector of fisheries in England, Arthur Berrington, warned, when it came to artificial propagation and breeding no one should expect special characteristics to be perpetuated 'as under changed conditions these peculiarities are apt to disappear'.

To Johnston, if scientists elsewhere accepted 'local' differentiation, adaptation, variation or disappearance of characteristics, then that approach ought to equally apply anywhere.

He posited that Tasmania had its own salmon variation, a home-grown one as it were. Fish deemed by some as an undoubted true salmon, but questioned by others, might just be a true analogue of *Salmo salar* with local characteristics.

He had culled his original seven possibilities to three races, or varieties, using existing classification tests: brown trout, growing much larger than in England; sea or salmon trout; and an 'intermediate form' attaining a larger size and freely entering salt water.

'If it be the true analogue of the English *Salar* it certainly has local characters which serve to distinguish it,' he said. And if classifiers persisted with their tests, 'we must recognise it for the time being by a local name.

'I propose for it the name of *Salar var Tasmanicus*, standing as a variety within the *Salmo trutta fario* species.'

26

END OF THE LINE

The christening of 'Tasmanian salmon' was well received. The Chief Justice, Sir Lambert Dobson, said Robert Johnston had gone back to first principles and abolished disputed terms and simply said: 'We have the salmon in a different variety.'

For the next few years, official handbooks for Tasmania referenced that its principal rivers 'are well stocked with various species of *Salmonidae*, *Salmo salar*, the salmon, etc'.

But the identity wars were not over. When Royal Society and museum curator Alex Morton noted some of Thomas Brady's hatchlings in the Salmon Ponds were of varying size, colour and characteristics, he wanted specimens sent to leading ichthyologists in Europe to verify they were true salmon. A large percentage agreed to the description of true salmon, but he was concerned that despite Brady's care and attention 'certain ova of trout had been sent out'.

Brady was outraged. Unless they looked like conger eels or kites, he said, then 'you did not receive one single ovum that was not taken out of the fish that we call *Salmo salar*, at any rate, fish that are exported and sold as such and that no man would have the hardihood to question when seen on the fishmonger's slab.'

The island had a fish called salmon, or at least a fish that was sold as salmon at home, he said. If that was acceptable in Ireland and England, was it not good enough for Tasmania?

Youl's quest for an unambiguous happy ending remained elusive, but New Year's Day 1891 gave him recognition no one could take away. The official list of honours included Queen Victoria's promotion of him to 'the most distinguished Order of St Michael and St George, to be ordinary members of the Second Class, or Knight Commanders of the most distinguished order'. At the beginning of his 80th year, this was a proud moment for a former sheep farmer once labelled a 'lunatic'. To be anointed Sir James Arndell Youl left him, a colleague said, 'quite overwhelmed'.

Youl pointed out to others that while several Tasmanians had been awarded the lower ranking of knight bachelor, 'this is the first time that the KCMG has been given to any Tasmanian colonist'. He joined a contingent of knights and officers at Paddington Station to train to Windsor Castle in March to receive their honours from Queen Victoria. Having spent so many freezing days on an everyday cushion to pack salmon ova to travel beyond the seas, he had the pleasure of watching the King of Arms to the Order carry a lush cushion bearing a badge, star and riband for the Queen's presentation.

Field said every Australasian angler would be delighted that the man who, 'above all others, has given trout – and for anything that is known to the contrary – salmon to their rivers, has been knighted'. It regarded Youl as 'always the moving spirit'.

Youl's spirits continued to be lifted by various reports read to him by his daughters. In 1893, Tasmanian agent-general Sir Edward Braddon told the Royal Society of Arts that salmon, salmon trout, and trout all 'abound'. Experts such as Sir Thomas Brady declared the salmon 'true',

—ENGLISH TROUT—
CAUGHT AT THE GREAT LAKE TASMANIA.
WEIGHT FROM 11 TO 17 Lbs.

Great Lake Trout captures were the talk of Britain. Photograph *Australian Town and Country Journal*, 1898. Courtesy National Library Australia.

or at least 'almost equivalent', matching fish sold as salmon in Britain. Braddon said it was 'at any rate a good imitation of that fish'.

In 1894 Youl heard William Senior, now angling editor of *Field*, tell an Imperial Institute dinner that the introduction of British fish had been a success, but whether the Tasmania fish 'remained salmon or trout, or whether they did not become a variety which in time would have to receive a distinct qualification ... suffice to say that the fish reached proportions which were never dream of in this country.'

Four years later, with Britain astonished by reports of giant trout catches in Tasmanian highlands, a large crowd attended an Imperial Institute exhibition of specimens, averaging 17½ pounds (8 kilograms), from Tasmania's Great Lake. The island's Fisheries Commission sent 17 'grand trout' in blocks of ice for exhibit in London, Liverpool, Manchester, Edinburgh, Glasgow and Dublin to attract tourists and anglers. After another lecture by Senior on Antipodean fish and fishing, Youl was enthusiastically applauded. He told attendees it was a proud moment for one who 'for 30 years or more had been called a lunatic for supposing the experiments could ever succeed'. He had said his motto through all the trials and torment had been 'what is worth doing is worth doing well.'

He was also pleased to hear the British Museum's George Boulenger declare that a fresh specimen from Waitaki, New Zealand, was unlike dozens he had inspected and was 'a true typical salmon, *S.Salar*, not a trout ... this is a salmon'. And German nobleman and experienced fisherman Baron Wulf von Bultzingsloewen reported a four-pounder caught on a visit to New Zealand was 'true grilse, not a trout; I am too old a salmon fisherman and have landed too many hundreds of grilse and salmon not to know the difference. There is not a shadow of doubt.'

Tasmania's desire for 'English salmon' had not expired, but shipments were abandoned after the *Kaikoura* in 1889. A visitor to the Salmon Ponds could now see more American than English fish being hatched, with ponds of quinnat and sockeye salmon, American brook trout, Californian rainbow trout.

In 1903, the Fisheries Commission revived plans for Atlantic salmon shipments, but declining government funding meant it did not eventuate. After more than half a century of hype and hope, effort and exasperation, the public was content with 'Tasmanian salmon'.

Others thought the island had given up too readily. The angler-author of a series in the *Australasian* said it was 'pretty lamentable' that the effort to acclimatise the finest sporting fish in the world was abandoned before prolonged experiment and patient investigation 'disclosed convincing reasons for the belief that the salmon will not live in Antipodean waters'. Without marking tags, selective netting or 'a hundred other means' to trace the progress from parr to smolt to grilse, informed judgments about reasons for their disappearance or the unsuitability of local waters were absent. 'An easier plan seems to have been adopted ... salmon parr were put in the Derwent, no salmon came back, as a result, ergo, local conditions are unfit.'

It seemed 'impossible' that English salmon could not succeed when conditions had not prevented brown trout and sea trout thriving, and the writer did not see evidence of a fair trial. Liberated fry deposited in the Plenty had to immediately run the gauntlet of trout, the 'big-headed cannibals ... that only feed on their own kin'. And as so few fry ever became a mature salmon, perhaps two per 1000, if most of the Derwent fish seemed to be neither pure salmon or pure sea trout then they could easily be hybrid progeny of the two.

Many in Tasmania were not fussed whether salmon had been gobbled by predators, become lost, headed to a new home, had developed into a local variant or engaged in hybrid breeding: 'Who knows?' was the common answer. As the *Australasian* concluded: 'The mystery of the salmon has not been solved, and there does not appear to be any probability that it will be.'

It was a mystery not resolved in Youl's lifetime. In the 1904 English summer, a cable went around the world: Sir James Arndell Youl, KCMG, had died at his residence in Clapham Park London on Sunday 5 June, aged 94, of 'senile decay' and bronchitis. He was buried with his first wife Eliza, and son Harold, at West Norwood cemetery in London. He left an estate of £159,853 (about £16million today, A$34 million) with a five-page will reflecting his lifelong appreciation of the value of a pound. With his three surviving sons, Charles, Cecil and Alfred, having already received land, he left his second wife, Dame Charlotte, £2000 (£207,000 today, A433,000) 'as a token of my love and esteem' and permission to occupy Waratah House for three months as long as she paid rates and taxes. 'She was assured of the fact that she had been adequately otherwise amply provided for was the sole reason for Sir James not making further provision for her.' Waratah House was to be sold with proceeds distributed to his five daughters.

Youl's death was reported throughout Australia, New Zealand, Britain, France and America. Newspapers listed the credits of 'a grand old man': successful pastoralist; secretary-treasurer of the Australian Association of Colonies'; agitator for colonial defences, mail services, local currency, emigration, an end to convict transportation and defender of the British Empire; founder of the Royal Colonial Institute, political agent for Tasmania; introducer of salmon and trout to Tasmania and New Zealand, gold medallist of La Societe d'Acclimatation, a knight.

Field said Youl had risen from being born in the 'bad old days' of convictism to become 'one of the great colonials, who, in a quiet way, did high service in laying the foundations of our Australian Commonwealth.'

As for the brotherhood of anglers, *Field* said 'to him, more than any other single individual, we owe the introduction of British *Salmonidae* to the rivers and lakes of New Zealand, Tasmania and Australia ... a distinguished work of acclimatisation.' While it was still an open question whether true salmon had been sustainably produced, Tasmania and New Zealand had become an angler's 'paradise', a remarkable achievement.

Between the *S. Curling* voyage of 1860, and *Kaikoura* in 1888, Youl shipped more than a million *Salmonidae*: about 876,000 salmon ova to Tasmania, and 569,000 to New Zealand; and more than 5000 brown trout, 24,000 sea trout or 'salmon trout', and about 7000 brook trout and 9000 lake char.

He outlived almost everyone who featured in his quest: Robert and William Ramsbottom, Edward Wilson, Francis Francis, Frank Buckland, Francis Day, Thomas Ashworth, Robert Buist, Louis Agassiz, William Denison, Gottlieb Boccius, John Shaw, Andrew Young, Morton Allport, William Archer, Robert Officer, Lauchlan Mackinnon, Victor Coste, Zepherin Gerbe, Joseph Remy, Antoine Gehin, John Gray, Francis Day, Abraham Bartlett, Charles Darwin, Charles Dickens, Thomas Huxley, Spencer Baird, Henry Moscrop, Alexander Black. His Irish ally Thomas Brady died three months later, Albert Gunther a decade on.

Youl died just a few weeks before a *Country Life Library of Sport* edition which would have much pleased him. George Boulenger, now cited by *Field* as the 'leading ichthyology authority in England', recalled Gunther's original rejection of a Tasmanian specimen as salmon and his insistence it was an aborted 'salmon trout'. Boulenger said Youl had 'protested at the time, with good reason, as the specimen, preserved in the British museum, is unquestionably a true salmon'.

'True salmon', Boulenger said, had been repeatedly caught in the estuaries and coasts of Tasmania and New Zealand, with 'the credit ... entirely due to James Youl, who planned and carried out the whole

James Youl, 1860. The year of his first salmon attempt. Photograph NS228/7/2/1, Tasmanian Archives.

Youl and family at Clapham Park. Photograph, PH30-1-6684.
Tasmanian Archives.

Youl and family croquet, Clapham
Park. Photograph, PH30-1-6683.
Tasmanian Archives.

The Day's Catch. Painting, Henry
Leonidas Rolfe (1847-1881).
Courtesy Bonhams, London.

Salmon in shop window, 1873. Watercolour, Frederick Walker.
Courtesy The Walters Art Museum.

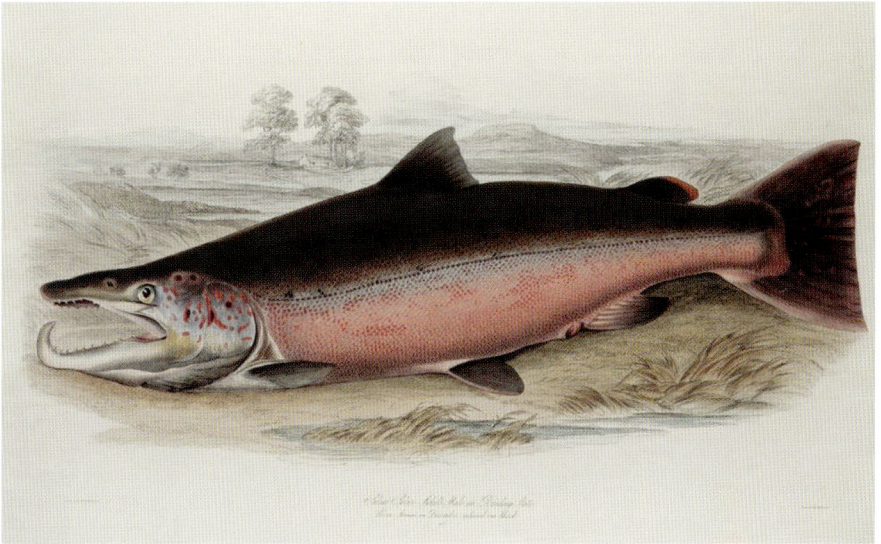

Salmon. Hand-coloured engraved plate, *Sketches from Nature* by William
Jardine, W.H. Lizars, Edinburgh,1839-1841. Courtesy Bonhams.

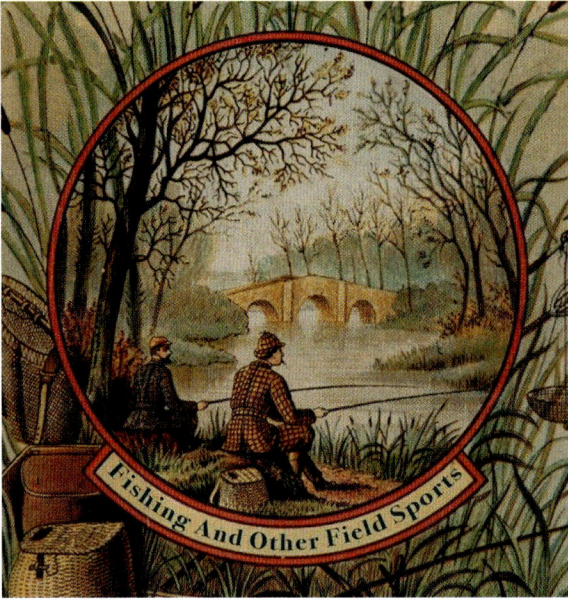

Old angling days. Catalogue illustration. Courtesy Sotherans, London.

Salmon and early flies, c. 1804. Original artist unknown. Published in *Rural Sports* by Rev William Daniel. Reproduced in *Field* Magazine, 2021.

Early explorers and settlers named native fish as Australian salmon because of perceived similarity to Atlantic salmon. Commonly called Eastern Australian salmon, blackback, cocky salmon, colonial salmon, and salmon trout, and in New Zealand as kahawai, but not related to Northern Hemisphere Salmonidae. Photograph, courtesy Department of Natural Resources and Environment, Tasmania.

Morton Allport. Self portrait, courtesy Allport Library and Museum of Fine Arts, State Libraries Tasmania.

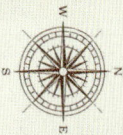

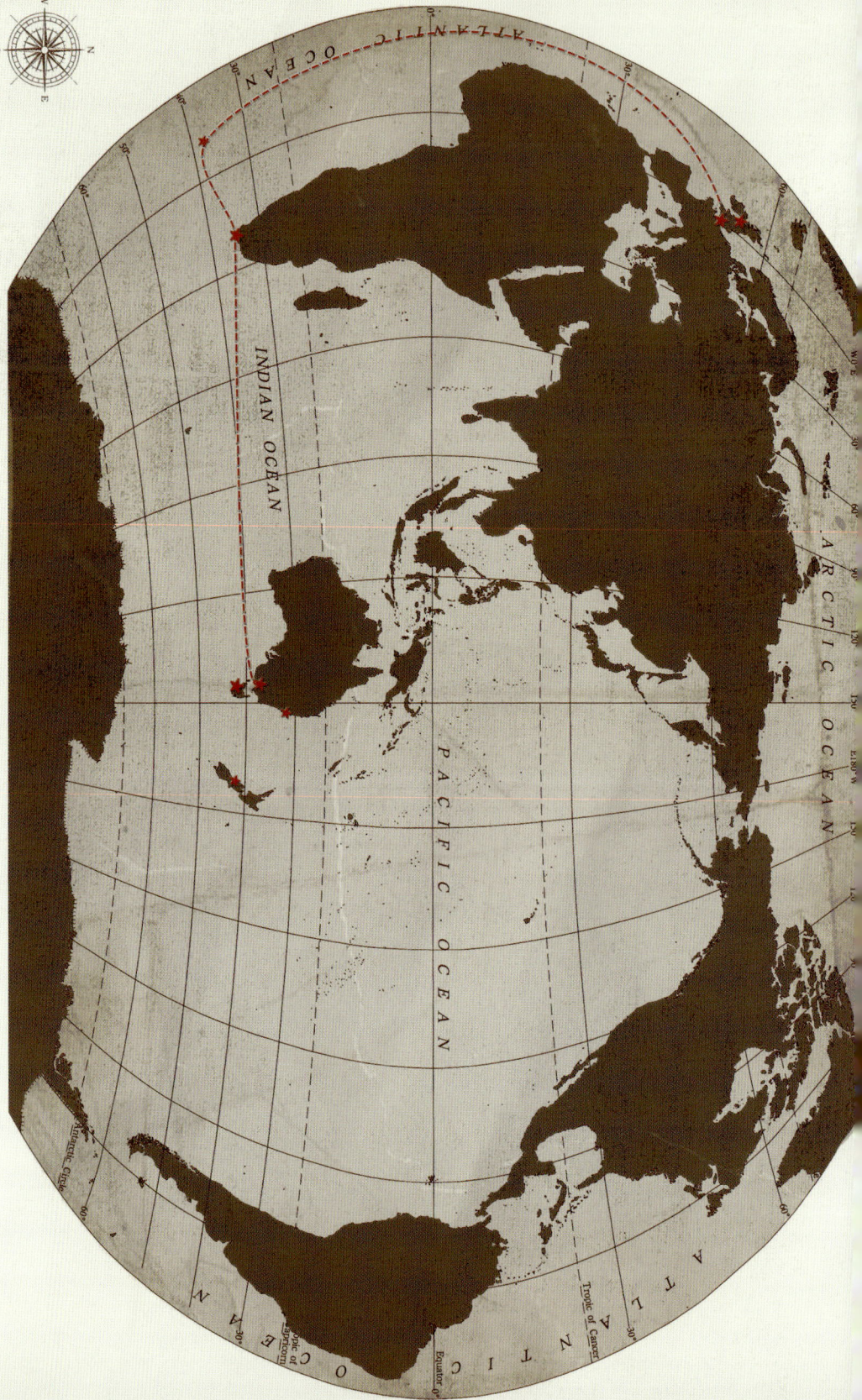

Left: Nineteenth century sailing route from England to Australia.

Huningue fish factoire. Colourised etching. Courtesy Exposition permanente Mémoire de Saumon, Association Petite Camargue Alsacienne, F - HAUT-RHIN-Saint-Louis.

Huningue incubation facility. Courtesy Exposition permanente Mémoire de Saumon, Association Petite Camargue Alsacienne, F - HAUT-RHIN-Saint-Louis.

| Ateliers pour l'incubation des oeufs | - Bâtiment Principal | *gravure de Kohl* |
| Eierbrutkanäle | - Central Gebäude | *Kupferstich von Kohl* |

Sandridge pier 1866. Hand-coloured engraving, original published in *Illustrated London News*. Courtesy Trowbridge, Claremont.

Waterfront Sullivan's Cove, Hobart, with view of ships and Salamanca warehouses, c. 1870. Photograph by J Backhouse Walker. Courtesy University of Tasmania Library Special and Rare Materials Collection.

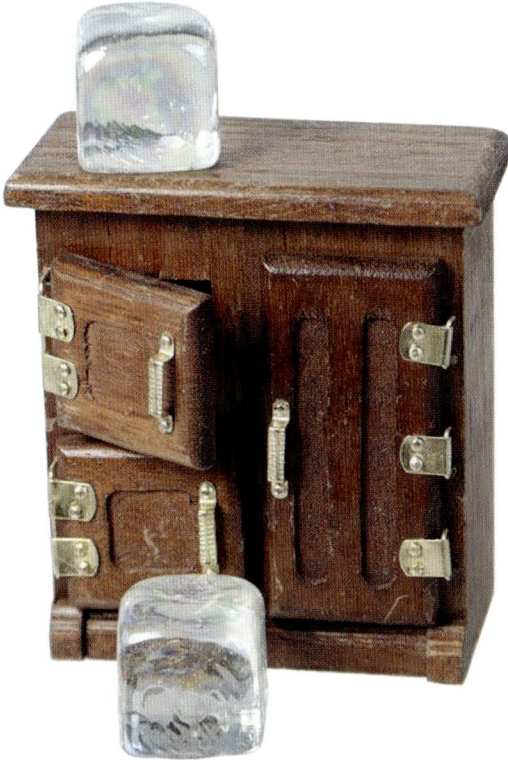

Early Victorian icebox. Alamy Stock Photo.

Youl ova box. Photograph by author at Australian Fly Fishing Museum, Clarendon, Tasmania.

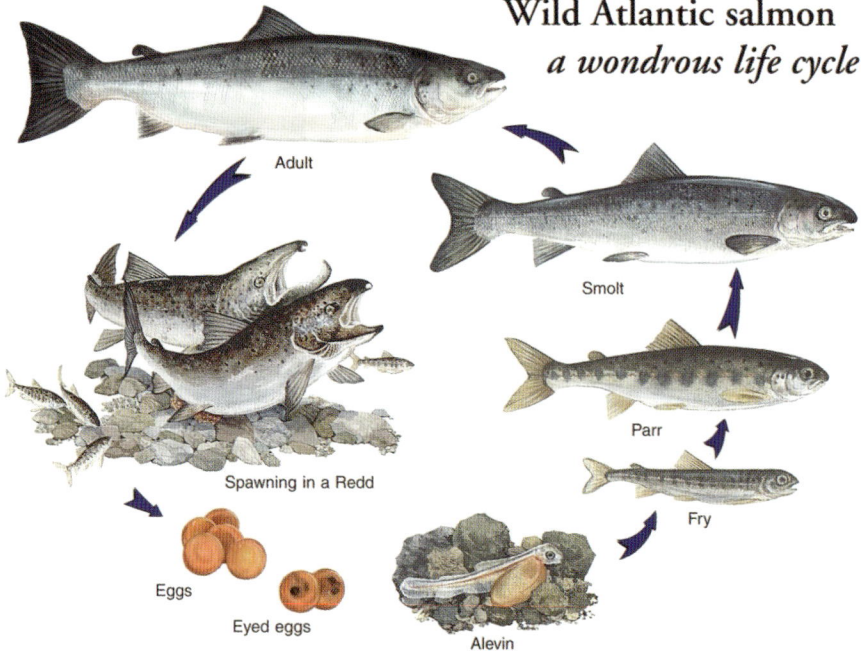

Wild Atlantic salmon
a wondrous life cycle

Adult

Smolt

Spawning in a Redd

Parr

Eggs

Fry

Eyed eggs

Alevin

Eggs to adult lifecycle. Illustration. Courtesy Atlantic Salmon
Federation, Montreal.

Alevin. Courtesy Paul Colvin, Happy Valley.
©paulcolvinphotography.com.

Salmon eggs. Courtesy US Fish and Wildlife Service.

Parr. Courtesy US Fish and Wildlife Service.

Smolt. Photograph Peter Steenstra. Courtesy US Fish and Wildlife Service.

Adult salmon. Photograph Ben Michelson. Courtesy Atlantic Salmon Federation.

SALMON PONDS NR. NEW NORFOLK.

Salmon Ponds, 1868. Photograph S Clifford. Courtesy
Libraries Tasmania.

THE SALMON PONDS, NEW NORFOLK, TASMANIA.

Salmon Ponds, 1874. Hand-coloured engraving, originally published
in Australasian Sketcher. Courtesy Antique Print
Map Room.

Ramsbottom cottage, Salmon
Ponds today. Photograph
by author.

Salmon Ponds today. Photograph by author.

Plenty River. Photograph by author.

French acclimatisation silver
medal, 1866.

French gold medal, 1869. Courtesy Noble Numismatics,
Melbourne.

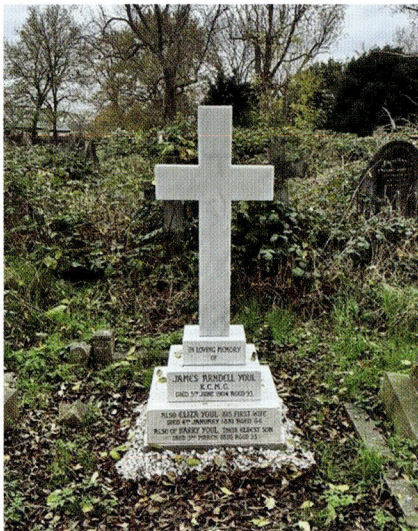

Youl headstone, renovated
2024, West Norwood Cemetery.
Courtesy Frank Youl.

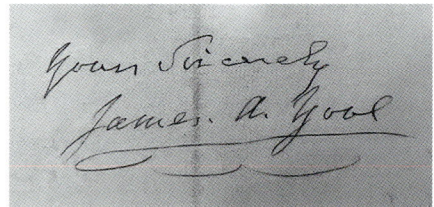

Youl signature.

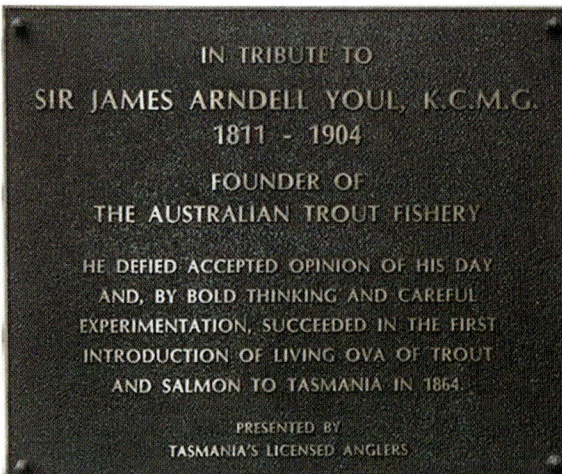

Youl plaque Salmon Ponds.
Photograph by author.

IN TRIBUTE TO
SIR JAMES ARNDELL YOUL, K.C.M.G.
1811 - 1904

FOUNDER OF
THE AUSTRALIAN TROUT FISHERY

HE DEFIED ACCEPTED OPINION OF HIS DAY
AND, BY BOLD THINKING AND CAREFUL
EXPERIMENTATION, SUCCEEDED IN THE FIRST
INTRODUCTION OF LIVING OVA OF TROUT
AND SALMON TO TASMANIA IN 1864.

PRESENTED BY
TASMANIA'S LICENSED ANGLERS

NZ Province of Otago silver cup to Youl, 1868. Photograph by author. Courtesy Andrew Youl.

Youl awarded Knight Commander of Order of St Michael and St George, 1891.

Youl portrait. Courtesy Cambridge University Library.

Leaping trout. Photograph Steven Ooi. Courtesy Inland Fisheries Service, Tasmania.

Leaping salmon. Photograph. Courtesy Charles Fleming, Devon.

Salmon pens Tasmania. Photograph Matthew Farrell. Courtesy *The Mercury*, Hobart.

scheme.' But most salmon disappeared, and the hopes of their taking to regular migration in the Antipodes had not been realised. Of the identity issue, Boulenger cited an old Swedish proverb, 'a dear child has many names'.

Two years on, Australian journalist-social reformer William Crooke maintained that 'the Derwent is at certain seasons literally alive with splendid salmonoids [*sic*]. Whatever variety they may be.' He suggested sending ova from Derwent salmonids back to England to hatch and be marked, and 'if they returned as true salmon there would be proof'. In his view, 'the salmon experiment ... is yet in its infancy.'

In the early twentieth century, with British blood and home ties diluted, most in Tasmania were content with a 'Tasmanicus' salmonid variation, or what the government regarded as 'at any rate a good imitation' of salmon, one readily sold and eaten in Britain as salmon. 'Tasmanian salmon' developed a cache, such that when fish shops in 1920s Melbourne were falsely selling imported fish from England, South Africa and Canada as 'Tasmanian salmon' Victoria had to give an undertaking it would cease.

In the 1940s, British Museum ichthyologist Dr Vladimir Tchernavin said Derwent specimens he had seen were true salmon: 'there could hardly be any doubt'. Through the 1950s the Tasmanian Fisheries Commission remained 'pretty confident' that some 'startling' catches were Atlantic salmon.

Today, some Tasmanian lakes are stocked with adult salmon, with a daily bag limit of five for licensed anglers, with only two permitted over 20 inches. Escapees from salmon fish farms are sometimes caught in coastal waters and estuaries. The Inland Fisheries Service's main focus is trout, especially the descendants of the original 'unwanted' brown trout, said to be the purest strains in the world and pursued and loved across Tasmania, mainland Australia and New Zealand. It maintains a steady supply of brown, rainbow and brook trout into lakes, dams and lagoons.

Today's lovers of trout and salmon are largely unaware that their angling and epicurean pleasure has its genesis in perhaps the most romantic, audacious and ambitious wildlife acclimatisation feat the world has seen. James Arndell Youl embellished the extent of some of his 'inspirations', and downplayed or did not always credit others who made valued contributions, or did not fully comprehend them, his determination and perseverance in delivering the first *Salmonidae* into the Southern Hemisphere for the first time, against prevailing knowledge, the tyrannies of distance and scientific scepticism, and sheer complexity, cannot be denied. In appalling weather, he obtained, packed, and dispatched nearly a million salmon eggs to Tasmania and half a million to New Zealand in a dozen shipments over two decades. His efforts were recognised in England, Ireland, France, Scotland, Germany, New Zealand and America, including imperial honours, gold and silver medals and silver plate. Such honours prompted, 40 years before his death, the *Mercury* to state that any community failing to recognise the achievements of its own was 'guilty of gross neglect and ingratitude ... Mr Youl has done his duty to the colony, it remains to be seen how the colony will do its duty as regards Mr Youl.'

Yet Youl never received significant Tasmanian recognition, no reward, and not the statue his closest supporters thought justified. At the Salmon Ponds, his name is included on a Tasmanian Anglers Hall of Fame board, and a small Hall of Fame plaque describes him as 'Founder of the Australian trout fishery'. It tells visitors: 'He defied accepted opinion of his day, and by bold thinking and careful experimentation, succeeded in the first introduction of living ova of trout and salmon to Tasmania in 1864.' In the Tasmanian Museum and Art Gallery's online exhibition of 100 objects exploring significant events and movements that have helped create the Tasmania we know today, the chapter on the *Norfolk's* historic salmon ova transport box makes no mention of Youl.

In London, Youl's simple headstone, which had long been in poor condition, was rectified by descendants in 2024 with support from the Fishmongers' Company, one of the oldest and most ancient Livery Companies of the City of London, 'in recognition for Sir James' connection to trout and freshwater species'.

While there is no sustainable wild Atlantic salmon population in Tasmania or New Zealand, '*Salmo Tasmanicus*' remains on the world register of marine species as 'a synonym for *Salmo salar*', the Atlantic salmon. More evident to the public, the Salmon Ponds at Plenty remains a popular working fish culture museum still contributing to trout and salmon production; Tasmania and New Zealand each bask as a famed 'paradise' of angling; and Tasmania has a substantial Atlantic salmon farming industry. All from those under the influence of salmon.

POSTSCRIPT

The 160 years since the first salmon, and 'happy accident' trout, were formally released to swim in the Southern Hemisphere for the first time, is a mere dot in the timeline of fish and human existence. But it is a dot which has seen remarkable and transformational changes in knowledge and challenges which have percolated since James Youl's day.

Some of the questions from his quest have been answered, or partially at least, while others have been elevated into existential debates and controversies. No other species has received as much scientific, commercial, technological and environmental attention.

Salmon is now better understood but still not completely with unresolved mysteries and an ever-evolving connection with us humans.

From a family tree which began some 15-20 million years ago when the ancestors of 'salmon' and 'trout' divided into two basic groups, with *Salmo* (Atlantic salmon, sea trout, brown trout) isolated in the North Atlantic and *Oncorhynchus* (Pacific salmon, rainbow trout) in the North Pacific, with a sub-family of *Salvelinus* (charr, brook trout, lake trout), the narrative has become both more magical and more complex.

Today's *Salmonidae* comprise 11 genera and about 70 species including salmon and trout, with many variations caused by local conditions and their 'plasticity'. Just as Tasmanians through the nineteenth and twentieth centuries were betwixt and between in identifying river trout, sea trout or 'salmon trout', and salmon, the genealogical maze continues: brown trout, for example, has more chromosomes than humans, so variable and adaptable with up to 50 separately named versions. 'Salmon' and 'trout' terms, as they were in the nineteenth century, are frequently interchanged, and not always exclusive to salt or fresh water. And the commercial manipulation of salmon has opened up whole new chapters.

As to the mystery of why so many salmon ova were shipped to Tasmania and New Zealand but so few were sustained or captured, Youl felt his best chances lay in replicating nature with its 'profusion' of salmon eggs. But only about 40,000 fry and smolt entered southern waters, not profusion by Nature's standards when so few salmon ova ever hatch and survive to reach adulthood, perhaps fewer than 30 out of 10,000.

For an introduced species to survive and be sustained, large and consistent additions are needed. Scientists call it 'propagule pressure'. Simply put, not enough salmon, now known to be poor colonisers, were introduced for long enough to build a sustainable breeding foundation. An insufficient run of survivors could not establish the patterns of migration, physiological changes, or 'memory' of navigation and smell, nor withstand cannibalistic trout, and predatory barracouta, sharks, seals and dolphins.

For those that did make it from the Salmon Ponds out of the Derwent, the 'lost at sea' notion has been elevated through oceanography and marine biology studies. Professor Thomas Quinn, of the University of Washington, and Dr Ken Lohmann, of the University of North Carolina, have shed light on how salmon, like other migratory birds and sea turtles, use their own magnetic map to guide them when heading to and from the sea. Without any inherited, instinctive, or established patterns, there was no natural 'map' in the Southern Hemisphere for the 'strangers' to turn to. They also lacked any real olfactory cues, or imprint of their home stream, vital to a salmon's return to its 'home' stream.

Explorers and sailors long utilised the global conveyor belt driving ocean waters continuously eastward, generally between 40° and 60° south latitude, but we now understand that between Tasmania and the South Island of New Zealand, the Antarctic Circumpolar Current diverges, with a split heading northwards through the Tasman Sea to meet the warmer waters of the Eastern Australian current sweeping south. Fish carried north on this route would probably perish in subtropical and nutrient-poor water or have to swim great distances against the current to rejoin the nutrient-rich current continuing east towards South America and the rich krill concentrations of the Scotia Sea, southeast of Cape Horn.

The southern marine world is very Antipodean, or opposite, for wild salmon. In the Northern Hemisphere, 'gyres' or circles of ocean currents formed by numerous landforms and different wind patterns, allow salmon to readily 'go with the flow' from fresh to sea water of the right temperature and food source and return to saltwater inlets and their freshwater tributaries of origin. The Southern Hemisphere denies such 'flow' to salmon, with huge distances separating African, Antarctic, Australian and South American continents. Scientists who studied salmon acclimatisation efforts in Tasmania, New Zealand and the Falklands concluded the absence of land masses and suitable ocean gyres largely explained the absence of adult salmon returning from the sea.

Today the issue is not whether salmon can live in the Southern Hemisphere. In fact they exist in the millions, just not as envisaged by Youl and others. The nineteenth century dream of 'artificial' fish propagation by 'seeding the water as earth' has taken on a new meaning.

Thanks to another sheep man, Youl's pioneering shipments can be seen as a formative chapter in a new world order. Norwegian Trygve Gjedrem, an agriculture student expected to follow his father as a sheep farmer, was asked in the 1970s by an animal genetics professor at the Agricultural University of Norway, Harald Skjervold, to help explore the application of animal breeding principles to fish. Gjedrem replied: 'But I don't know anything about salmon.' To which Skjervold responded: 'None of us do!'

But the professor knew that, over many centuries man had manipulated wild animals into domesticated sheep, chickens, cows, pigs, but 70% of the globe, the sea, remained largely untapped. Against doomsayers saying they were fools – 'This is ludicrous! We don't want anything to do with that rubbish' – the Norwegians' extensive genetic studies led to marine paddocks producing fast growing, cost-effective, protein-rich, consumer-friendly salmon. In a blink, wild salmon became domestic salmon.

Fast, fat salmon meant fast, fat rewards. Salmon farming quickly changed the world's diet, especially after the Norwegians persuaded the Japanese to adopt salmon instead of tuna, creating a sushi and sashimi tsunami. Salmon was on its way to becoming a food consumed by billions of people every day, with the biblical exhortation to 'have dominion over the fish' taken to massive levels.

After Skjervold visited Tasmania and affirmed its waters were also ripe for salmonid farming, a joint venture between the Tasmanian Fisheries Development Authority and Norwegian investors led to the first commercial sale of recognised Tasmanian Atlantic salmon in 1987. While Youl and others envisaged 'gold nuggets', they could not have imagined salmon farming becoming easily the island's largest individual primary industry, with a production value of $1.4 billion a year – exceeding dairy, wool, wine and fruit etc – and about 70 licensed inland hatcheries and marine farms producing 90% of Australia's Atlantic salmon consumption.

The island's local producers, Tassal, Petuna and Huon, are all enmeshed in a $60 billion global salmon industry, having been brought by Brazil's JSB meat-processing giant, Canada's Cooke Seafood, and New Zealand's Sea Lord. Salmon farming is the fastest growing food system in the world, with billions of dollars being spent by companies and countries riding the salmon's back to food and fortune in the quest to cost-effectively feed more people with essential protein.

Nineteenth century industrial age faith, or 'madness' as some maintained, in artificial propagation to replenish and grow fish stocks as an affordable food supply for growing populations has morphed into a twenty-first century biotechnology age faith in fish-farming to feed the world's population as it rises to 10 billion around 2050, driving a 50% increase in demand for food. Today 95% of salmon consumption is farmed fish, not wild.

Hundreds of millions of ova are hatched and raised to smolt stage, and then progressively moved into and through marine tanks and floating pens or 'paddocks' to be fattened until they are ready to be processed. But the dramatic growth and expansion has come with echoes of nineteenth century concern. The 'leave nature alone' versus 'man can do better than nature' debates of the old world are echoed today as policy makers, scientists, corporations, anglers, environmentalists and consumers contest how much latitude should be given to 'improving' nature.

The story of how the beef, chicken, lamb and pork on our plates evolved over centuries from wild aurochs, jungle fowl, mouflon and boar is largely forgotten. Issues around the domestication, manipulation and industrialisation of animals and birds for food, including cruelty, soil degradation, water and air pollution, waste, and genetic engineering, has largely been seen off or accepted.

The historically admired salmon is another story in the narrative of 'progress'. The transformation from wild species into mechanised domesticity and burgeoning food industry has taken place in less than a lifetime, coinciding with a growing world needing to feed itself but also recognising the shame that humans account for 0.01 percent of all life but have destroyed more than 80% of all wild mammals and half the plants.

In the wild, the survival of wild salmon, a keystone species in the world's marine web, has become endangered, even extinct, in many areas due to human interference. Its sensitivity to warming waters also sees salmon regarded as a 'canary in the gold mine'. As King Charles, patron of the Atlantic Salmon Trust, put it, 'when all is well with the salmon, all is well with the world'.

Farmed salmon, more than any other species, finds itself amid a maelstrom of debates and controversies around food efficiency, economics and security; environmental and species sustainability; technological overreach and bioethics; globalisation and geo-political tensions; corporate greed and transparency; politics and regulation. In Tasmania, salmon continues to speak: desired in the nineteenth century to help define a 'new' Tasmania, then as a welcome new farming industry in the twentieth century, now whether it can reflect and reinforce Tasmania's twenty-first century reputation of having some of the world's cleanest air and water, with a unique connection to the environment. In Tasmania, salmon continues to speak: desired in the nineteenth century to help define a 'new' Tasmania, then as a welcome new farming industry in the twentieth century, now whether it can reflect and reinforce Tasmania's twenty-first century reputation of having some of the world's cleanest air and water, with a unique connection to the environment.

In a still evolving industry, the salmon rush has seen many salmon producers pay insufficient regard to the fact they operate with more than a production license. Their social license, the ability to operate and profit in public waterways, hinges on a supportive public and consumer trust, more than animal or land farming, and such trust is increasingly fragile.

Unsurprisingly, the industry has opened itself up to criticism by prize-winning author-journalists, scientists and environmentalists. In their prize-winning 2021 work, *The New Fish*, Norwegians Simen Saetre and Kjetil Ostlimarine challenged their salmon-rich country and the consequences of *Homo sapiens* too quickly becoming homo deus, a God-like species wanting to rule over the dominions of nature and species. Since then, Richard Flanagan has written *Toxic: the Rotting Underbelly of the Tasmanian Salmon Industry* and Americans Douglas Frantz and Catherine Collins *Salmon Wars, The Dark Underbelly of our Favourite Fish*.

The 'underbelly' theme is that today's marine fish-farming is the antithesis to a planet striving to survive by becoming cleaner, healthier and more sustainable for all species. Cited crimes include the need for up to 1.5 kilograms of wild fishmeal and oil to grow one kilogram of salmon; a carbon footprint per pound higher than chicken; the negative impact of nutrient outflows on marine species, biodiversity and public drinking water; escaped fish impacting wild salmon through disease and crossbreeding leading to genetic dilution of pure salmon and more 'hybrids'.

It is not a million miles from the nineteenth-century question of Henry David Thoreau, 'who hears the fishes when they cry?', or Charles Dickens' observation that 'it might be worthwhile, sometimes, to inquire what Nature is, and how men work to change her, and whether in the enforced distortions so produced it is not natural to be unnatural.'

In an echo of nineteenth century 'artificial fish' wars, more extreme critics protest against ungodly 'man-made fakes ... that should not be called a salmon', of 'Frankenfish', and a McWorld of homogenisation through 'caged super-chickens of the sea' and 'marine battery hens'. But what is undeniable is that salmon has evolved into two distinct biologies, which Professor Mart Gross, an evolutionary biologist at the University of Toronto, says are declining wild (*Salmo salar*) and rising domesticated (*Salmo domesticus*) species. 'When a fish ... is removed from its wild habitat and domesticated over generations for human consumption it changes the fish and our perception of it.' Gross sees domesticated salmon 'as different from wild salmon as dogs are from wolves.' Noted American fish author Mark Kurlansky agrees: 'Gastronomically, a wild salmon and a farmed salmon have as much in common as a side of wild boar has to pork chops.' Even the stock of famed Alaskan 'wild salmon' is aided by 30 hatcheries releasing two billion juvenile salmon annually, showing that even 'wild' does not always mean a complete absence of human management.

The 'Atlantic salmon' or 'Tasmanian Atlantic salmon' in our supermarkets, sushi chains and restaurants is not the salmon of yesteryear. It is not a fish which leaps and migrates on specific routes for generations, has a unique lifecycle as it moves between fresh and salt water, inheriting or developing the magical ability to find its way home to breed.

Salmon 'domesticus' runs in a different lane, with an unnatural habitat, lifecycle, feeding and breeding habits. It is retained and 'enhanced' in tanks and pens, where it loses its hearing, relies on a diet of other marine life or artificial pigments to give consumers their favoured orange-pink flesh colour. Genetic manipulation sees true males removed from the bloodstock to ensure better quality and returns. Life support comes from automatic feeding, chemicals, disease protection and antibiotics. Predators are kept at bay through fences, underwater explosives, and cap shooting – and sometimes echoes of nineteenth-century tactics, such as Tassal employees in 2024 shooting more than 50 great cormorants as part of a shoot-and-scare tactic in a failed bid to deter the native seabirds from entering salmon-farming pens.

In-shore farming operations draw an easy spotlight, with more visibility and controversy around issues of feed, disease, sea lice and parasites, toxic algae, marine diversity, water quality, and waste. Fuelled by environmental opponents, some of whom are multi-million-dollar operations needing their own sustenance, visibility easily move from local to national to global audiences. At remote Macquarie Harbour on the west coast of Tasmania, where the salmon industry has operated quietly for 40 years, the rare Maugean skate, a species of ray carrying World Heritage value for its link to dinosaurs, was seen to be at risk of becoming extinct. While the industry and some researchers argued no one understood the precise cause or timing of declining numbers – be it salmon farming, treated wastewater, freshwater from hydro power, rising ocean temperatures or mining runoffs – the 'guilty' finger was pointed at the salmon. Multi-million dollar government and industry investment in oxygenation and advanced monitoring for two years has seen an improvement in skate numbers, but 'stop salmon expansion' and 'shut down salmon' calls elevated in the shadows of an Australian election.

The salmon heat rose further when it was revealed a bacterium outbreak, which the industry said was elevated by warmer water temperature but presented no risk to humans, led to at least a million salmon deaths at Tasmanian fish farms, with dumping at landfill sites and rendering plants, and reports of fatty chunks of fish washing up on beaches in the Huon Valley and on Bruny Island. Notwithstanding that not long before the Australian chicken industry, which routinely culls more than 12 million unwanted male chicks each year, similarly had to cull more than 1.5 million chickens after an outbreak of bird flu, critics pounced on salmon farming as 'an animal welfare nightmare'.

Hollywood superstar Leonardo Dicaprio, surfing legend Mick Fanning, and film director Justin Kurzel harnessed social media, and T-shirts, to promote 'Vote Salmon Out!' an 'Eating Salmon? Killing Tasmania' messages, and local conservation and Indigenous groups began lobbying for 'international embarrassment' by having the United Nations' World Heritage Committee send a monitoring mission. While his Environment Minister struggled to juggle the scientific, environmental and World Heritage arguments in the Macquarie Harbour controversy, the Prime Minister promised to 'protect jobs' and legislate for protection of 'sustainable farming'.

Controversies draw attention, but community resolution of long-term and complex issues is rarely helped when opposing sides seek 'weaponising' opportunities. Some of the salmon farming criticism has been well justified, and acknowledged by some in the industry, but some has been toxic. On the other hand, the industry would be unwise to continue proclaiming the election results, where most overtly anti-salmon candidates did not succeed, as an unambiguous 'wholehearted' backing of current modus operandi. Governments, too, need to consider the efficacy of environmental policies and regualtion.

Nevertheless, attention can produce action, fuelling new ways to produce salmon – at sea or on land – more efficiently, more sustainably for fish and other species, more transparently and more aligned with community expectations. Around the world, profit and social license concerns are driving billions of dollars into research, technology and innovation, forging new frontiers of what is viable on land and at sea. As Professor Daniel Benetti, a respected American aquaculturist, has observed: 'Self-interest is driving improvements … not because people are nice. They're there to make money. But sustainability and profitability come hand in hand'

Governments, such as Norway, and big companies are seeking to 'modernise' aquaculture policy and operations in the quest for better and more sustainable outcomes for all stakeholders, and to deal with geopolitical tensions. All in some way driving towards 100% closed containment systems, either land-based or in-ocean, as the growing mantra.

Companies are developing methods to feed farmed salmon with necessary nutrients and fatty acids without needing to harvest wild anchovies, sardines, herring, and capelin. More plant-based ingredients, such as algae are being trialled, along with protein cultivated from microbes in soybean processing waste, animal waste and by-products, and insect larvae. Feed-based treatment against sea lice infestation is being explored as a replacement for chemical or mechanical treatments.

New offshore technologies will see more operations move out to sea, offering the potential of enhanced environmental conditions, colder and more natural ocean currents, avoidance of sea lice and toxic algae, and better management of excess feed and waste. In Tasmania, a co-operative government-industry-research operation, Blue Economy, is undertaking a three-year trial of two advanced fish farming pens about 12 kilometres offshore from Burnie in Bass Strait. In New Zealand, the government has approved a Malaysian-backed 'Blue Endeavour' development of the country's first offshore King Salmon farm seven kilometers at sea in northern Marlborough, South Island.

Other concepts include massive submersible rigs, drawing inspiration from the oil and gas industry, and perhaps co-located with ocean wind farms. Ocean Arks Technology in Chile is developing a 170-metre 'super yacht' allowing fish cages to be mobile to better withstand storms and waves up to eight metres, avoid algal blooms, and ensure optimal levels of oxygen and temperature.

Some companies doubt the economic or ecological sustainability of sea farming. Riverence – America's largest trout producer, based in Idaho – is one of several companies developing land-based farm models offering the potential for egg-to-market facilities under one roof, with closed or re-circulating technologies to deal with water, temperatures, salinity, oxygen, currents, lighting, CO2, waste, energy, feed and waste, and remove the need for pesticides and antibiotics.

In Tasmania, owner JBS is investing $110million into Huon's land-based Whale Point farm, to allow seven million salmon per year to be farmed longer in recirculating tanks before being released into sea-based pens for shorter periods, with treated waste being used as fertiliser for local cherry farms and in pet food production.

Rising geo-political tensions is leading more countries to be concerned about food security. Singapore, which has to import 90% of its food and is thus vulnerable to any food supply disruption, international conflict, or disease outbreaks, wants more local aquaculture, which saw plans for an eight-storey fish farming facility. China wants to guarantee its food supplies (and elevate its economic power) through a range of projects: massive fish farms on Gaotang Island and in the Gobi Desert; billion-dollar deep-sea farming cages, the height of a 20-storey building, 200 kilometres out at sea; floating fish farm vessels; aquaculture industrial parks and cities; and 50 'demonstration zones for mariculture' in Shandong region. (The country has also blurred salmon identity, permitting locally farmed rainbow trout to be labelled and sold as salmon).

The United States similarly wants to reduce the country's reliance on seafood imports – about 95% of its Atlantic salmon consumption is imported – with development of land-based aquaculture using self-contained recirculating aquaculture systems.

Governments, industry, and the public share the universal love for salmon, but also share a need for salmon food to be produced with more responsibility, transparency, and accountability. That will bring broader and deeper meaning to aspirations of 'responsibly farmed salmon' (Tassal), producing food which is 'sustainably and ethically sourced, grown and processed' (Petuna) and evidencing 'the meticulous care we show our fish and our environment' (Huon).

Much has changed since the nineteenth century, but we have modern echoes of the question of Henry David Thoreau, 'who hears the fishes when they cry?', or Charles Dickens' observation that 'it might be worthwhile, sometimes, to inquire what Nature is, and how men work to change her, and whether in the enforced distortions so produced it is not natural to be unnatural.' The aspirations of firms like AquaBounty, an American bio-technology company, to combine 'the goodness of nature with the power of science and technology to give more people reliable access to...salmon' is not so far removed from the 'artificial fish propagation' of the industrial age.

What a future pen of history writes is yet to be revealed, but perhaps in the end we will see the same questions from this book's beginning: questions about Man and Nature, attitudes and latitudes, identity and place. As salmon goes, be it wild or domestic, so do we.

ACKNOWLEDGEMENTS

More than one family member and friend raised their eyebrows during the research and writing of this book. 'A whole book about salmon, really!?' Often followed by a quizzical or sceptical 'Why?'

While at times even I wondered if I was as 'mad' as James Arndell Youl was initially described, my response, was simple: the tale intrigued me and I thought it deserved to be better known, especially in light of the salmon controversies of today, which remind me of one of my favourite quotes, from American author William Faulkner in 1950: 'The past is never dead. It's not even past'.

The task took me on a biographical journey of a fish, a man, and an island. From paleolithic cave images to twenty-first century genetic sequencing images; peasant anglers to billion-dollar fish farming companies; industrial revolution to biotechnology revolution; acclimatisation to climate change. Along the way I 'met' men and women whose interest in salmon in some way has shaped lives not just in Tasmania, but around the world.

While book-writing, like angling is largely a very personal and silent pursuit, this was a voyage of discovery that needed many people to come alongside to encourage, educate and guide. My thanks go to all those who did so, and hope they recognise it's impossible to personally reference everyone.

It was energising to have respected authors, academics and scientists around the world willingly provided material, insights, and navigation tips, or just plain encouragement. They include Dr Anders Halverson, of Colorado and Mark Kurlansky, of New York, whose salmon and trout books truly engage, educate and challenge; Richard Le Boucher, Research Director at Singapore's Temasek Life Sciences Laboratory, for his input on the history and future of salmon; Chris Newton, of Gloucestershire, for his enthusiasm and sharing his fine trout book. Dr Noel Wilkins, of Galway, whose early interest helped get the project off the ground; Dr Jesse Trushenski, Chief Science Officer of Reverence in Idaho, for her observations on salmon farming; Professors Thomas Quinn of the University of Washington, Tony Ricciardi of McGill University, and Julie Lockwood, of Rutgers University, for their expertise on salmon; Grant Jesse, secretary of World of Water, for research assistance; Professor Marianne Lien, social anthropologist at the University of Oslo, for her thoughts on Tasmanian Atlantic salmon and the politics of belonging; and Australian salmon and angling pioneers Mick Hortle and Mick Hall for their guidance.

Staff at numerous bodies deserve thanks, including: State Library Tasmania; Allport Library, Tasmanian Heritage and Archives Office; National Library of Australia; National Trust, Tasmania; Australian Fly Fishing Museum, Tasmania; Salmon Ponds, Tasmania; Clitheroe Civic Society, Lancashire; Bibliothèque Nationale de France; Societe Nationale de la Nature, Paris; Petite Camargue Alsacienne salmon museum; Scientific Society, Arcachon, France; Royal Commonwealth Society Collections at Cambridge University Library; Special Collections at Manchester University; Victoria University of Wellington; Oughterard Heritage Group; Galway City Museum; Galway National Library.

Digital doors which afforded invaluable access to original resources include: Biodiversity Heritage Library, the world's largest digital library for archival and contemporary biodiversity literature; the Biblioteque Nationale de France Gallica; Google Scholar; Forgotten Books; Project Gutenberg; JStor; National Library of Australia's Trove and access to

British and Australian newspapers and periodicals; Library of Congress; Royal Society for the Arts; Cambridge University Library for records of the General Association of the Australian Colonies, Royal Colonial Institute and Royal Commonwealth Society; Royal Botanic Gardens (Melbourne) Baron von Mueller correspondence project;.

Such agencies provided remarkable access to original writings and correspondence of key players in the evolving salmon tale: Izaak Walton, Ludwig Jacobi, Joseph Remy, Antoine Gehin, Charles Darwin, Frank Buckland, Francis Francis, Thomas Ashworth, John Shaw, Louis Agassiz, Jules Haime, Victor Coste, Livingstone Stone; Spencer Baird; George Marsh, Humphry Davy; Thomas Garnett; Robert Knox; Robert Buist, Thomas Stoddart; Gottlieb Boccius; Albert Gunther, Francis Day, Ferdinand von Mueller.

On James Youl, I drew on his original correspondence, collected from numerous newspapers and periodicals, including the *Times*, *Morning Post* and *Field*, (UK) and the *Argus*, *Age*, Hobart *Mercury*, Launceston *Examiner*, *Cornwall Chronicle*) (Australia). Also, his correspondence with the Royal Society of Tasmania, Salmon Commission, General Association of the Australian Colonies, Edward Wilson, Lachlan Mackinnon, Morton Allport, William Archer, Samuel Wilson, Francis Francis, William Lauder Lindsay. The extensive writings of Allport and the annual reports of the Salmon Commission (1862-1887) and papers of the Royal Society of Tasmania were also invaluable, as were the nineteenth-century journals and correspondence of Alexander Black and William and Robert Ramsbottom.

I also gratefully acknowledge Catherine Pearce's generous sharing of her unpublished 2020 manuscript, *Antipodes, a Nineteenth Century Quest to bring Servant Girls and Salmon to Australia*; Bob Starling's 2013 family history research on Reverend John Youl and his descendants; the 1980 family history compiled by John and Robin Youl, *Youl the Immigrant*, 1801-1882; Philip Blake's biography of Rev John Youl, and the interest of James Youl's descendants: Frank, Andrew and the late Michael Youl.

On Tasmanian history, I drew on the University of Tasmania's *Companion to Tasmanian History*; Henry Button's *Flotsam and Jetsam, Floating Fragments of Life in England and Tasmania*; LL Robson's *History of Tasmania*; various works by Professor Stefan Petrow at the University of Tasmania; James Fenton's *A History of Tasmania*; WC Piguenit's *The Salmon Ponds and Vicinity New Norfolk*; John West's *History of Tasmania*; James Boyce's *Van Diemen's Land*; Katie Febey's University of Tasmania thesis *Who'll come a Waltzing Matilda with me? Stock Theft and colonial Relations in Van Diemen's Land*; Walch's Tasmanian Almanac; Heritage Tasmania; Tasmanian Parliament; Historical Records of Australia; plus, the published work of colonial newspaper editors and reporters.

On Robert and William Ramsbottom, I acknowledge the Ramsbottom family of Edmonton, Canada, especially Josh, 3x great grandson of William Ramsbottom, for generously allowing access to family history material; Shirley Penman, archivist at Clitheroe Civic Society for unwavering interest and assistance; Fred Higham, chairman of Ribblesdale Anglers association and Ian Appleton, of the Whitewell Fishing Association for their knowledge of Ribble-Hodder.

I gratefully thank Dr Ryan Wilkinson, Inland Fisheries Service of Tasmania; Luke Martin and Dr John Whittington, Salmon Tasmania; and Frank and Andrew Youl for tangibly supporting this history project. Likewise, I again thank publisher David Tenenbaum at Melbourne Books, and applaud his commitment to supporting nonfiction Australian history. The editing support of Georgia Cooper and the design skills of Holly Lambert are much appreciated.

On a personal level, the book would have been impossible without the forbearance of my wife Maureen, and an inheritance of historical curiosity and perseverance from my mother and late father.. Additional encouragement when swimming against the tide came from my three children - Shannan, Elise and Clayton - and Michael Gawenda, Andrew Rule, Adam Courtenay, Geoff Slattery, Ray and Chris Henderson, Muriel Reddy, Neil and Judy Travers. I particularly thank Geoffrey Blainey, a true gentleman and scholar, for his special friendship and support and generous offer to pen a foreword. I also thank all those friends who, through company or caffeine, have helped turn the pages of life.

SELECT BIBLIOGRAPHY

Allport, Morton

> *Account of the Introduction of Salmon Ova into Tasmania*, Papers and proceedings of the Royal Society of Tasmania (RST) Hobart 1864
>
> *On the Natural Enemies of the Salmon in Tasmania*, RST, Hobart 1864
>
> *The Attempt to introduce Salmon Ova by the Beautiful Star*, RST, Hobart, 1864
>
> *Letterbooks, Allport Library and Museum of Fine Arts, Libraries Tasmania, 1864-1878*
>
> *Report of the late successful Experiment for the Introduction of Salmon Ova and Sea Trout Ova to Tasmania*, RST, Hobart 1866
>
> *Brief History of the Introduction of Salmon and other Salmonidae to the waters of Tasmania*, Proceedings Zoological Society of London (ZSL), London, 1870
>
> *The Salmon Trout*, RST, Hobart 1871
>
> *Observations on the Progress of the Salmon Experiment*, RST, Hobart 1871
>
> *Further Notes on the Salmon Experiment*, RST, 1874
>
> *Some further Notes on the Introduction of the Salmon into Tasmanian Waters*, RST, 1874
>
> *On the Introduction of Salmon to the Waters of* Tasmania, Proceedings ZSL, London, 1874
>
> *Further Notes on the Salmon Experiment*, RST, 1875
>
> *Present stage of the Salmon Experiment*, RST, 1877

Anderson, Warwick, *Climate of Opinion: Acclimatization in Nineteenth Century France and England*; Victorian Studies, Vol. 35, No. 2, 1992, Indiana University Press, Bloomington (USA), 1992

Ashworth, Thomas and Ashworth, Edmund, *A Treatise on the Propagation of Salmon and Other Fish*, E.H. King, London, 1853

Ashworth, Edmund, *Propagation of Salmon: Remarks on the Artificial Propagation of Salmon and Some Account of the Experiment at Stormontfield*, Hasler & Company, Bolton, 1875

Baggini, Julian, *How the World Eats; A Global Food Philosophy*, Granta, London, 2024

Behnke, Bob, *The Ramsbottom Family- Fish Culturists*, American Fly Fisher, Vol. 21, #1, Manchester (US), 1995

Bertram, James, *Harvest of the Sea*, J. Murray, London, 1865

Black, Alexander, *Journal S.Courling* (sic) 1860, RST, Hobart, 1860

Boccius, Gottlieb

> *A Treatise on the Management of Fresh-water Fish, with a view to making them a source of profit to landed proprietors*, London, J Van Voorst, 1841

Fish in Rivers and Streams; a treatise on the management of fish in fresh waters, by artificial spawning, breeding and rearing: showing also the cause of the depletion of all rivers and streams, J Van Voorst, London, 1848

Bompas, George, *The Life of Frank Buckland*, Smith, Elder, & Co., London, 1885

Bonow, Madeleine; Olsen, Håkan; Svanberg, Ingvar, *Historical Aquaculture in Northern Europe*; Sodertorn University, Fleminsberg (Sweden) 2016

Buckland, Francis, *Fish Hatching*, Tinsley Bros., London 1863

Buist, Robert, *The Stormontfield Piscicultural Experiments*, 1853-1866, Edmonston and Douglas, Edinburgh, 1866

Clements, John, *Salmon at the Antipodes: a history and review of trout, salmon and char and introduced coarse fish in Australasia*, J Clements, Ballarat, 1988

Coates, Peter, *Salmon*, Reaktion Books, London, 2006

Day, Francis, *Salmo Salar and Salmo Ferox in Tasmania*, Nature, November 1885, Macmillan and Co., London, 1885

Dickens, Charles, *Salmon*, All Year Round, London, 1861

Dunfield, RW., *The Atlantic Salmon in the History of North America*, Canada Department of Fisheries and Oceans, Ottawa, 1985

Dunn, Bob, *Angling in Australia: its history and writings*, David Ell Press, Balmain, 1991

Francis, Francis, *Fish-Culture, a Practical Guide to the Modern System of Breeding and Rearing Fish*, Routledge, London, 1863.

Fry, William, *A Complete Treatise on Artificial Fish-breeding: including the reports on the subject made to the French academy and the French government; and particulars of the discovery as pursued in England*, D. Appleton and Co., New York, 1854

Garlick, Dr Theodatus, *A Treatise on the Artificial Propagation of Certain Kinds of Fish*, A.O. Moore, New York, 1858

Girling, Richard, *The Man Who Ate the Zoo: Frank Buckland, forgotten hero of Natural History*; Chatto and Windus, London, 2016

Grant, Jesse and Casey, A., *A Study of the Chronological Dates in World Aquaculture History from 2800BC*, World of Water.uk.org, updated Dec. 2024.

Halverson, Anders, *Entirely Synthetic Fish – How Rainbow Trout Beguiled America and Overran the World*, Yale University Press, New Haven, 2010

Harwood, J. Keith, *The Ramsbottoms: Pisciculturists, Tackle Manufacturers, and Fly Dressers*, American Fly Fisher, Vol. 27, #4, Manchester (US), 2001

Haxo, J. Dr., *On the Artificial Fertilization of Fish eggs and their Hatching: via processes created by Messrs. Rémy and Géhin, from Bresse.* De l'Imprimerie de veuve Gley, Epinal, 1852

Jardine, William, *Illustrations of British Salmonidae with Descriptions*, Jardine, Edinburgh, 1839, 1841

Johnston, Robert, *General and Critical Observations on the Fishes of Tasmania*, Royal Commission report, Hobart, 1882

Kinsey, Darin, *Seeding the Water as Earth*, Environmental History, University of Chicago, Chicago, July 2006,

Kurlansky, Mark, *Salmon, A Fish, the Earth, and the History of a Common Fate*, One World, London, 2020

Lever, Christopher, *They Dined on Eland: The Story of the Acclimatisation Societies*. Quiller Press, London, 1992

Lichatowich, Jim

  Salmon Without Rivers: A History of the Pacific Salmon Crisis, Island Press, Washington, 1999

  Salmon, People and Place, Oregon State University Press, Corvallis (US) 2013

Lien, Marianne, 'King of Fish' or 'Feral Peril, *Tasmanian Atlantic salmon and the Politics of Belonging*; Environment and Planning D: Society and Space, 2005

Lindsay, WL, *Salmon Acclimatisation in New Zealand*, Royal Society New Zealand, 1873

Mackenzie, Sir Francis, *Brief and Practical Instructions for the Breeding of Salmon and other Fish Artificially*, Taylor and Francis, London, 1841

Marshall, A.J., *Darwin and Huxley in Tasmania*, Hodder and Stoughton, London, 1970

McDowall, Robert, *Gamekeepers for the Nation: the story of New Zealand's acclimatisation societies*, Canterbury University Press, Christchurch, 1994

McDowall, Robert, *Trout in New Zealand Water*, Wetland Press, Wellington, 1984

Minard, Pete

  All Things Harmless, Useful, and Ornamental: Environmental Transformation through Species Acclimatization, from Colonial Australia to the World, North Carolina Press, Chapel Hill (US) 2019

  Salmonid Aclimatisation in Colonial Victoria, Environment and History, Vol 21, University of Chicago, Chicago, 2015.

Moscrop. E.H., *Correspondence relative to the introduction of salmon and trout at the Antipodes*, printed Wightman and Co., London, 1879

Nash, Colin, *The History of Aquaculture*; Wiley Blackwell, Iowa, 2011

Newton, Chris, *The Trout's Tale, The Fish that Conquered an Empire*, Medlar, Ellesmere (UK), 2013

Nichols, Arthur, *The Acclimatisation of the Salmonidae at the Antipodes, its history and results*, London, S. Low, Marston, Searle, & Rivington 1882

Osborne, Michael, *Acclimatizing the World: A History of the Paradigmatic Colonial Science*, Nature and Empire: Science and the Colonial Enterprise Vol. 15, University of Chicago Press, Chicago, 2000

Owen, James, *Trout*, Reaktion Books, London, 2012

Pauly, Daniel, *Darwin's Fishes, Encyclopedia of Ichthyology, Ecology and Evolution*, Cambridge University Press, Cambridge, 2007

Pearce, Catherine, *Antipodes, a Nineteenth Century Quest to bring Servant Girls and Salmon to Australia,* unpublished, Launceston, 2020

Piguenit, WC, *The Salmon Ponds and Vicinity New Norfolk*, Piguenit, Hobart, 1867

Ramsbottom, Robert, *The Salmon and its Artificial Propagation*; Simpkin, Marshall, London, 1854.

Ramsbottom, William

 Diary, Beautiful Star voyage 1862, RST, Hobart, 1862

 Report on the Late Experiment on the Introduction of salmon Ova into Tasmania,
 Salmon Commission annual report, Hobart, 1862

Ritchie, Jack, *The Australian trout, its Introduction and Acclimatisation in Victorian Waters*,
 Victorian Fly Fishers association, Melbourne, 1988

Robson, LL, *History of Tasmania*, Vols. 1, 2, Melbourne University Press, Melbourne,
 1983, 1990

Rolls, Eric, *They All Ran Wild, The Story of Pests on the Land in Australia*, Angus and
 Robertson, London, 1977

Roosevelt, RB, and Green, Seth, *Fish Hatching and Fish Catching*; Union Advertiser,
 Rochester, 1879

Russel, Alexander, *The Salmon*, Edmonston and Douglas, Edinburgh. 1864

Seager, Phillip S, *A Concise History of the Acclimatisation of the Salmonidae in Tasmania*,
 RST, Hobart, 1888

Saetre, Simen, and Ostli, Kjetil Ostli, *The New Fish, the Global History of Salmon Farming*;
 Cappelen Damm, Oslo, 2021

Saville-Kent, William

 On the Acclimatisation of the Salmon in Tasmanian Waters, 1887 RST, Hobart, 1887

 Nature, Macmillan and Co., London, 1885

 Superintendent and Inspector of Fisheries reports, Parliament Tasmania, Hobart,
 1885-1887

 The Naturalist in Australia, Chapman and Hall, London, 1897

Scholes, David, *Trout Quest*, Stevens Publishing, Launceston, 2005

Seager, Philip, *Concise History of the Acclimatisation of Salmonidae;* RST, Hobart 1888

Senior, William, *Travel and Trout at the Antipodes*, Chatto and Windus, London, 1880

Shaw, John, *Experimental Observations on the Development and Growth of Salmon Fry, etc.*
 John Shaw. Edinburgh. 1840.

Stewart, Leslie

 A History of Migratory Salmon Acclimatisation Experiments in Parts of the Southern
 Hemisphere and the Possible Effects of Oceanic Currents and Gyres Upon Their
 Outcome; Advances in Marine Biology, CSIRO, Cleveland, (Aust) November 1980

 Salmon in New Zealand, with special reference to Falkland Islands, Ministry of
 Overseas Development, London, 1975

Stone, Livingstone, *Salmon Breeding*, Transactions of the American Fisheries Society,
 Transactions of the American Fisheries Society, Volume 21, Issue 1, January 1892,
 Bethesda (USA)

Tchernavin, V., *The Origin of Salmon, Its Ancestry, Marine or Freshwater*, Salmon and
 Trout Magazine, Number 95, Salmon and Trout Association, Torrington (UK) 1939

Towle, Jerry, *Livingstone Stone and the Transformation of Californian Fisheries*, American
 Society for Environmental History, Oxford University Press, Oxford, 2000

Trushenski, Dr Jesse

    From Johnny Fish-Seed to Hatching-Bashing to Shaping the Shoal of Aquatic Stakeholders; Fisheries, American Fisheries Society, Bethesda (USA), August 2020,

    Land-Based Aquaculture for Atlantic Salmon, Salmonland.org, 2022

    Turchini, Giovanni; and Glencross, Brett, Thoughts for the Future of Aquacultural Nutrition, North American Journal of Aquaculture, 2019.

Walker, Jean, *Origins of the Tasmanian Trout: an account of the Salmon Ponds and the first introduction of salmon and trout to Tasmania in 1864,* Inland Fisheries, Hobart, 1988

Walton, Izaak; Cotton, Charles, *The Compleat Angler,* (orig. 1676), Penguin Random House, Sydney, 1998

Wigan, Michael, *The Salmon, The Extraordinary Story of the King of Fish*; William Collins, London, 2013

West, John, *History of Tasmania,* Henry Dowling, Launceston (Aust.) 1852

Wilkins, Noel

    Ponds, Passes and Parcs, Aquaculture in Victorian Ireland, Glendale Press, Dublin, 1989

    A Species of Delusion: the Inspectors of Irish Fisheries 1819-2019; Institute of Public Administration, Dublin, 2020

Wilson, Sir Samuel, *Salmon at the Antipodes; being an account of the successful introduction of salmon and trout into Australian waters,* London, E. Stanford 1879

Young, Andrew

    Fitzgibbon, Edward (Epherema), Book of the Salmon, Longman, Brown, Green and Longmans, London, 1850

    The Natural History and Habits of the Salmon; With Reasons for the Decline of the Fisheries. Longman, Brown, Green, and Longmans, London, 1854.

Yarrell, William, *A History of British Fishes,* J. Van Voorst, London, 1836

Useful websites included: UN Food and Agricultural Association, www.fao.org; the World Wildlife Fund, www.worldwildlife.org; Atlanticsalmontrust.org; wildfish. org; WorldofWater.uk.org; www.globalsalmoninitiative.org; Global Biodiversity Information Forum, www.gbif.org; salmonland.org; www.wildsalmon.org; livingoceans.org; Aquaculture Stewardship Council, www.asc-aqua.org; the Tasmanian Department of Natural Resources and Environment, www.nre.tas.gov. au; Salmon Tasmania, www.salmontasmania.au; Inland Fisheries Tasmania, www. ifs.tas.gov.au; National Oceanic and Atmospheric Organisation www.fisheries.noaa. gov; Atlantic Salmon Foundation, Canada, www.asf.ca; Atlantic Salmon Trust, www. atlanticsalmontrust.org

THE AUTHOR

Steve Harris is a former editor, editor-in-chief and publisher at *The Age*, *Sunday Age* and *Herald Sun*. He is a life member of the Melbourne Press Club, a Knight Fellow at Stanford University, and in 2024 was made a Member of the Order of Australia for significant contributions to print journalism.

A fourth-generation Tasmanian, he is the author of *Solomon's Noose*, the true story of a young convict who became Queen Victoria's longest-serving hangman; *The Lost Boys of Mr Dickens*, about the British Empire's first juvenile prison in Van Diemen's Land; and *The Prince and the Assassin*, about Australia's first royal tour and act of political terrorism.